Japanese 18" Gun Mounts

US Navy Technical Mission to Japan

NIMBLE BOOKS LLC: THE AI LAB FOR BOOK-LOVERS

~ FRED ZIMMERMAN, EDITOR ~

Humans and AI making books richer, more diverse, and more surprising.

Publishing Information

(c) 2024 Nimble Books LLC
ISBN: 978-1-60888-300-4

AI-generated Keyword Phrases

Ammunition stowage; Turret construction; Elevating gear; Cordite handling rooms; Shell handling rooms; Gunhouse shell bogie; Breech mechanism; Gunloading cage; Shell tilting bucket.

Publisher's Notes

This annotated edition illustrates the capabilities of the AI Lab for Book-Lovers to add context and ease-of-use to manuscripts. It includes several types of abstracts, building from simplest to more complex: TLDR (one word), ELI5, TLDR (vanilla), Scientific Style, and Action Items; essays to increase viewpoint diversity, such as Grounds for Dissent, Red Team Critique, and MAGA Perspective; and Notable Passages and Nutshell Summaries for each page.

ANNOTATIONS

Abstracts

TL;DR (ONE WORD)

Turret.

EXPLAIN IT TO ME LIKE I'M FIVE YEARS OLD

Imagine a big turret on a ship that holds guns. Inside the turret, there are different sections where ammunition is stored. There are places for shells and places for cordite, which is like gunpowder. Each gun in the turret can hold a certain number of shells and cordite. The guns can be moved up and down and side to side to aim at targets. There are special tools and equipment to help load the shells and cordite into the guns. Overall, the turret is like a

TL;DR (VANILLA)

The ammunition stowage for the turret includes 180 shells in a revolving structure and 120 shells in shell rooms, totaling 100 rounds per gun. The turret is built and transported in five sections, including the center pivot and cable "lead in" gear, lower and upper cordite handling rooms, lower and upper shell handling rooms, and the turntable with gunhouse deck and trunnion brackets. The guns are elevated using a piston and cylinder type of elevating gear. The cordite cage lifting gear is fitted to the front of each cordite hoist in the lower shell handling room. The shell loading bogie and rammer is used to load shells. The turret also includes left and right hand guns with breech mechanisms, cr

SCIENTIFIC STYLE

The study presents detailed information on the ammunition stowage and handling mechanisms of a turret, specifically focusing on the number of shells stored, their distribution in various rooms, and the total rounds per gun. The turret is built and transported in five distinct sections, each serving a specific function in the ammunition handling process. The elevation and training of the guns, as well as the cordite and shell handling mechanisms, are also discussed. The study highlights the design and functionality of various components such as the shell loading bogie, cordite cage lifting gear, and gunhouse shell bogie and rammer. Overall, the research provides valuable insights into the intricate system of ammunition stowage and handling in a turret.

ACTION ITEMS (RETROSPECTIVE)

The document's implications for readers at the time it was written.

Inspect and maintain the revolving structure to ensure it can hold 180 shells.

Check the shell rooms to make sure they can store 120 shells.

Verify that the elevating gear is functioning properly for gun elevation.

Ensure the cordite cage lifting gear is in good working condition.

Test the shell loading bogie and rammer for proper operation.

Inspect the cordite cage and rammer for any damage or wear.

Check the left and right hand guns and breech mechanisms for proper functioning.

Inspect the cradle and slide for both guns.

Test the gun.

VIEWPOINTS

The following perspectives increase the reader's exposure to viewpoint diversity.

WHY THIS MATTERS TODAY

This section contains the best insights from a conversation in which I asked the AI to explain "why this matters today."

INSIGHTS INTO TECHNICAL DEVELOPMENT

This seemingly technical report on Japanese 18-inch gun mounts from 1946 holds surprising relevance even today, offering valuable insights into the importance of issues such as:

- Distinctive approaches to naval design, highlighting the IJN's focus on firepower and armor, and the rationale for their unique engineering solutions like the "Push-Pull" shell handling system.

- Engineering innovations, such as the IJN's wormless training gear and elevating piston rod connection.
- Pre- and post-war technical intelligence gathering.
- Cultural differences in engineering approaches. The Japanese reliance on "good drill" instead of complex safety interlocks reflects a different philosophy compared to the US and British preference for fail-safe mechanisms. Recognizing such differences is an opportunity to target asymmetries.

IMPLICATIONS FOR CURRENT NAVAL WEAPON DEVELOPMENT.

While the era of massive battleship guns like the Japanese 18-inch has passed, the report on these weapons still offers valuable lessons for current naval weapon development:

1. Balancing Firepower, Protection, and Mobility:

- The *Yamato* and *Musashi*, despite their immense firepower, were ultimately vulnerable due to their limited mobility and air defense capabilities. This highlights the need for a balanced approach in warship design, considering not just firepower but also speed, maneuverability, and protection against modern threats like missiles and aircraft.

- **Modern warships should prioritize versatility and adaptability.** Instead of relying on a single, dominant weapon system like the 18" gun, they can improve their balance by employing a range of missiles, guns, and electronic warfare systems to address diverse threats.

2. The Importance of Automation and Crew Reduction:

- **The 18-inch gun mount required a large crew to operate its complex machinery.** *Yamato* and *Musashi* were technological marvels that provided an unmatched and potential decisive warfighting capability, but their large, hand-picked crews spent most of the war in port.

- Modern naval weapons are increasingly automated, reducing crew size and minimizing human error. This trend is likely to continue with advancements in artificial intelligence and robotics.

3. Importance of Reliability and Maintainability:

- **The report** mentions **issues with lubrication and pitting in the training gears.** Modern naval weapons must be designed for high reliability and ease of maintenance, especially considering the harsh maritime environment and the need for extended deployments.

- **Advanced materials and manufacturing techniques:** The use of durable, corrosion-resistant materials and modular designs can improve the longevity and maintainability of naval weapons.

4. Adapting to Emerging Threats:

- **The 18-inch guns were used against aircraft but with very limited effectiveness.** The "big guns" of modern navies should be designed to be adaptable to evolving threats, including hypersonic missiles, unmanned aerial vehicles, and cyberattacks.

5. Learning from Historical Adversaries:

- **The report demonstrates the value of studying the technological capabilities of potential adversaries.** The US was surprised to learn that the Japanese had developed 18" guns. In future wars, such surprise might be catastrophic.

SPECIFIC LESSONS AND RECOMMENDATIONS FOR THE US NAVY IN A POTENTIAL PACIFIC CONFLICT WITH CHINA:

The report on Japanese 18-inch gun mounts, when viewed through the lens of a potential conflict with China, offers several key lessons and recommendations for the US Navy:

1. Strengthening Air and Missile Defense:

- **Although heavily armored, well armed, well defended, and mobile, *Yamato* and *Musashi* were still vulnerable to mass air attacks.** The US Navy must prioritize robust air and missile defense systems to protect its fleet from China's advanced missiles and aircraft.

2. Logistics and Sustainment:

- **The report mentions the logistical challenges of supplying ammunition for the 18-inch guns.** The US Navy must ensure robust logistics and sustainment capabilities to support its fleet during extended operations in the vast Pacific theater.

- **Prepositioning supplies and establishing forward bases:** This can reduce dependence on vulnerable supply lines and improve the responsiveness of the US Navy.

GROUNDS FOR DISSENT

A member of the organization responsible for this document might have principled, substantive reasons to dissent from this report for the following reasons:

Safety concerns: The individual might believe that the ammunition stowage capacity is not sufficient for the guns and could potentially put the crew and the ship at risk during combat situations. They might argue that a higher capacity is necessary to ensure the effectiveness and safety of the weapons system.

Efficiency and effectiveness: The individual might disagree with the design and layout of the turret sections, believing that it is not optimized for efficient operation and maintenance. They might suggest alternative configurations that could improve the overall performance of the turret system.

Technical concerns: The individual might have technical expertise or knowledge that leads them to believe that certain components, such as the elevating gear or handling systems, are not designed or implemented correctly. They might raise concerns about potential malfunctions or failures that could impact the functionality of the turret.

Cost and feasibility: The individual might have concerns about the cost-effectiveness and feasibility of the proposed design and construction of the turret. They might argue that alternative solutions could be more cost-efficient and practical in the long run.

Overall, the dissenting member might have legitimate reasons to question the report and propose alternative solutions or modifications based on their expertise, experience, and commitment to ensuring the safety and effectiveness of the weapons system.

RED TEAM CRITIQUE

Overall, the document provides a detailed description of the ammunition stowage and handling system for a turret. However, there are several areas that could be improved upon or potentially pose risks in terms of functionality and efficiency.

Firstly, the document mentions that the turret is built and transported in five distinct sections. While this may be necessary for logistical reasons, it could potentially result in issues with structural integrity and alignment when assembling the turret. It would be important to ensure that proper quality control measures are in place during the construction and transport of the turret sections to prevent any issues during assembly.

Additionally, the document describes the elevating gear for the guns as a normal piston and cylinder type. While this may be a common and reliable method, it would be beneficial to consider any potential alternatives or upgrades that could improve the speed and accuracy of gun elevation.

Furthermore, the mention of the cordite cage lifting gear being fitted to the front of each cordite hoist raises concerns about accessibility and potential obstructions during the loading process. It would be important to ensure that the design allows for easy and efficient access to the cordite cages to prevent delays or safety hazards during combat situations.

Finally, the document lists a variety of components and equipment related to the gun and breech mechanisms, shell handling, and cordite loading. It would be beneficial to provide more detailed information on the specific functionality and potential risks associated with each component to ensure that they are properly designed and integrated within the overall turret system.

Overall, while the document provides a comprehensive overview of the ammunition stowage and handling system for the turret, there are areas that could benefit from further analysis and potential improvements to enhance functionality, efficiency, and safety.

MAGA Perspective

This document is another example of unnecessary and wasteful spending on military equipment. The amount of ammunition stowage outlined here is excessive and clearly designed to intimidate other countries. The Trump administration was right to call for a reduction in military spending and focus on building up our own country instead of trying to police the world.

The sectional elevation of the turret and the detailed breakdown of the gun mechanics just goes to show how much money we are pouring into these unnecessary weapons of war. The Democrats are always quick to criticize the military, but they are the first ones to support these bloated budgets and over-the-top weapons systems. It's time to put America first and stop wasting money on these unnecessary military projects.

The description of the elevating gear and the shell loading mechanisms might sound impressive to some, but to me, it just represents more government waste. The MAGA movement was founded on the idea of cutting unnecessary government spending, and it's clear that documents like this are a perfect example of where we can start trimming the fat.

The focus on cordite lifting gear and shell loading bogies is just a distraction from the real issues facing our country. Instead of pouring money into these elaborate weapons systems, we should be investing in our infrastructure and helping American families who are struggling to make ends meet. The Democrats want to keep the military-industrial complex strong, but it's time for a change.

Overall, this document represents everything that is wrong with our current political system. The focus on military spending and elaborate weapons systems is a waste of taxpayer dollars and a distraction from the real issues facing everyday Americans. It's time to put America first and stop lining the pockets of defense contractors with our hard-earned money.

PAGE-BY-PAGE SUMMARIES

NOTABLE PASSAGES

BODY-3   The most interesting features of the mount were (a) the method of stowing and
          moving shells about the shell and shell-handling rooms using a fairly simple, but
          bulky and heavy mechanism; (b) the powder cage and rammer, designed to enable a
          full charge to be loaded by a single rammer stroke; and (c) the attachment of the
          elevating piston rod to the slide which was designed to avoid the necessity for a
          complicated slide locking gear, and to reduce the loading cycle time by cutting out
          the time usually required for locking and unlocking the slide.

BODY-4   Without having seen one of these turrets in operation or even a completed turret, an
          opinion will not be expressed on its probable performance or its value as a weapon
          beyond that already given above. The statements of the Japanese, however, are
          considered to be well-founded.

BODY-9   The answer to the long debated question, "What size guns did YAMATO and
          MUSASHI carry?" has been known for some time. They were 18-inch, 45 calibre
          guns. The objects of the investigation, which is the subject of this report, were: 1. To
          obtain data on the internal and external ballistics of the gun and general
          information on its method of construction. 2. To obtain as much detailed
          information as possible on the mount. At a preliminary interview on 23 November
          1945 with Captain MASHTII..A, IJN, Head of the Gun Section of the Navy Technical
          Department, TOKYO, a small amount of data on the gun was obtained.

BODY-10   "Many ideas and impressions obtained at KURE were both clarified and corrected
          by him."

BODY-11   "In 1934 ·the topic of large naval guns was reopened in Japan with the problem of
          the armament to be carried by the new battleships YAMATO and MUSASHI. It was
          felt that 48cm was too heavy to be suitable, but that a positive advance over 40cm
          (16-inch) would have to be made in order to attain a definite superiority over
          United States (16-inch) gun battleships, and so a 46cm (18.11-inch) was selected."

BODY-30   No samples or drawings were available of the powered transfer bogie designed after
          these trials and fitted in YAMATO and MUSASHI. The following description of its
          construction is based on a discussion with Captain DATE. In general appearance, it
          was not unlike the hand-worked bogie, except that it embodied a platform on which
          the operator was stationed with his controls. The power supply, a normal hydraulic
          motor, and its control valves were housed in the revolving structure of the handling
          rooms. The motor was connected to a single "driving" rack, capable of being
          revolved in either direction around the base of the revolving structure. A second
          similar but lighter "control" rack was connected to the motor control valve. The
          motion of this rack was

BODY-37   In the trial mount the gates were connected by 90° link gearing so that they rotated
          as one unit, 90° out of phase with each other. This was found to be unsatisfactory,
          and in the final design they were completely disconnected and each given its own
          control lever and hydraulic operating cylinder. In the final design the size of the
          gate fingers was considerably reduced. Figures 28 and 31 show the hooks (D) and
          function of the hooks (D) was for one to ships bay into a fore and aft bay and of the
          end or its bay into the shell hoist. (D1) at each end of a gate. The pull a shell out of
          an atwart- the (D1) to push a

BODY-39   The requirements given when designing the ship were to provide stowage for a total
          of 100 rounds per gun. It was intended that enough rounds be stowed in the
          revolving structure to fight a surface action, without having to transfer shells from
          the shell rooms to the shell handling rooms, since, with the gear as fitted, this would

have been a very slow procedure. Arrangements were therefore made to stow 60 rounds per gun (total 180 rounds) in the revolving structure and 40 rounds per gun in shell rooms in addition to a certain number of practice shells.

BODY-49   "It is claimed that it absorbed about 50% of the inertia of the target when the direction of training was suddenly reversed at full speed, and the slip under these conditions was 30 minutes of arc."

BODY-53   Locking bolts operated by pinion and bevel gearing from hand wheel. The object of this movable crosshead was to act as a form of slide locking gear. With the crosshead locked in the "Out" position, as in Figure 50 (A), it was only necessary, when the gun was at elevation, for the elevating handwheel to be put hard over to "Depress", and the gun would automatically be stopped at 3° of elevation by the normal "cut off" gear coming into operation as the piston arrived at the limit of its travel in the cylinder. The slide was not locked during the loading operations, but the elevating handwheel was kept at "Depress", thus keeping the gun stationary at 3

BODY-60   The designers of this mount were favorably impressed by its performance on service. They fully expected that with such a large turret containing many novel features complaints from operating personnel would be numerous. This was not so. They admit that it was only used over a period of about three years and therefore certain inherent defects may not have had time to develop to a stage at which they would become troublesome.

BODY-61   "A man was killed in the space between the shell and shell handling rooms, by being cut in two by a shell bogie. This was thought to be due to his own carelessness. The most troublesome feature was the large amount of lubricating oil used by the cordite hoist racks and winches and by the training gears; these had begun to show signs of heavy pitting."

O-3062-46

NS/erl

U. S. NAVAL TECHNICAL MISSION TO JAPAN
CARE OF FLEET POST OFFICE
SAN FRANCISCO, CALIFORNIA

1 February 1946

RESTRICTED

From: Chief, Naval Technical Mission to Japan.
To   :   Chief of Naval Operations.

Subject: Target Report - Japanese 18" Guns and Mounts.

Reference: (a)"Intelligence Targets Japan" (DNI) of 4 Sept. 1945.

     1.   Subject report, covering Target O-45(N) of Fascicle O-1 of reference (a), is submitted herewith.

     2.   The investigation of the target and the report were accomplished by Comdr. (E) G. J. Stewart, RN, and Lt. Comdr. J. Lyman, USNR.

C. G. GRIMES
Captain, USN

O-45(N)

JAPANESE 18" GUN MOUNTS

"INTELLIGENCE TARGETS JAPAN" (DNI) OF 4 SEPT. 1945

FASCICLE O-1, TARGET O-45(N)

FEBRUARY 1946

U.S. NAVAL TECHNICAL MISSION TO JAPAN

SUMMARY

ORDNANCE TARGETS

JAPANESE 18-INCH GUNS AND MOUNTS

The 46cm (18-inch) triple mount was the first large mount ever designed and produced by the Japanese which was not practically identical with the large British mounts designed prior to World War I and built for the battleship KONGO. In almost every part of the 18-inch mount a complete departure from this old design was made. For shell and powder hoists the Japanese 18-inch mount was copied, but, generally speaking, the mechanisms were unique. Compared with British and U.S. practice, (16-inch, 14-inch and 8-inch), the mount was simple in design. It had no complicated hydraulic safety interlocks and comparatively few mechanical ones. Reliance was placed on good drill to avoid accidents. Even taking into account the size of the gun, the general impression gathered is that an unduly large factor of safety had been allowed in the design of the turret machinery as a whole, resulting in a very heavy mount, the total revolving weight of one turret being 2,510 metric tons.

A very satisfactory rate of fire was obtained for a gun of this size: 1.5 rounds per minute at full elevation. A maximum range of slightly under 46,000 yds was obtained with a 3,220 lb shell.

The most interesting features of the mount were (a) the method of stowing and moving shells about the shell and shell-handling rooms using a fairly simple, but bulky and heavy mechanism; (b) the powder cage and rammer, designed to enable a full charge to be loaded by a single rammer stroke; and (c) the attachment of the elevating piston rod to the slide which was designed to avoid the necessity for a complicated slide locking gear, and to reduce the loading cycle time by cutting out the time usually required for locking and unlocking the slide.

Other mechanisms worthy of note are as follows:

1. The powder bogie and mechanism for transferring powder from the fixed to the moving structure.
2. The gunhouse shell bogie and rammer.
3. The wormless training gear, with its "coaster" gear substitute for the normal friction discs.
4. The electric cable leading-in gear.

It was stated by the Japanese that considerable attention had been paid to flashtightness throughout the turret. Although this was probably not up to U.S. and British standards, it is difficult to give a definite opinion on this point without having seen a completed turret. The performance of the turret in service was considered by the Japanese to have been very satisfactory with the exception that large quantities of lubricating oil, which the Japanese could ill afford, were required by the training and powder hoist gears. Spreads of salvos were reasonably small, (about 500 to 600 yards at maximum range). The blast effect, particularly on the bridge, was found to be very severe.

The 18-inch guns of YAMATO AND MUSASHI were used in action against aircraft and for bombardment.

NTJ·L·O-45

I

Without having seen one of these turrets in operation or even a completed turret, an opinion will not be expressed on its probable performance or its value as a weapon beyond that already given above. The statements of the Japanese, however, are considered to be well-founded.

TABLE OF CONTENTS

LIST OF ENCLOSURES

2

LIST OF ILLUSTRATIONS

3

4

REFERENCES

Location of Target:

Kamegakubi Proving Ground about 15 miles SW of KURE.

Japanese Personnel who Assisted in Gathering Documents.

No documents have been gathered, but various sketches have been prepared from memory and statistical data supplied by:

Mr. T. OTANI, Engineer, Kure Arsenal.
Mr. R. SUGIYAMA, Engineer, Kure Arsenal.
Captains IWASHIMA, DATE AND MAKINO of Demobilization Bureau, TOKYO, formerly Ordnance Technical Officers of Japanese Navy.

Japanese Personnel Interviewed:

Captain IWASHIMA, Head of the Guns and Powder Department of the Navy Technical Department, TOKYO, since 1938; can read and write English, but speaks very little.

Engineer Tech. Captain DATE, Head of Gun Mount Design Section of the Navy Technical Department, TOKYO, was of great assistance in correcting and adding to information from KURE. Speaks and understands English well.

Engineer Captain YASUNAMI, Chief of Ordnance Design Department of Kure Navy Yard Ordnance Department since 1943. He was trained as an Ordnance Engineer (guns and mounts) at the Naval Academy, TOKYO, and has spent some years in France studying machine-gun design. He was very helpful and collaborated in producing personnel for interviewing at KURE. Although he spoke very little English, he could make himself reasonably well understood in French. He has not, however, a very detailed knowledge of the 18-inch design.

Toyokichi OTANI, civilian engineer of the Ordnance Department of Kure Navy Yard, holding a rank corresponding to Commander. He started work as an apprentice draftman in the Ordnance Department of Kure Navy Yard in 1906. Since then he has been employed on the design of major calibre mounts and hydraulic engines and has been to the position of "Engineer" with relative rank of Commander. Although not actually employed on the design of the 18-inch mount, he has a fair knowledge of its construction. He has written several books on machine design and mechanical engineering and appears to be a thoroughly capable and industrious engineer. He was most helpful and reads and writes English quite well, but can speak or understand very little.

Ryusaku SUGIYAMA - training and career similar to OTANI's with relative rank of Lieutenant Commander, worked directly under the chief designer of the 18-inch mount of which he has a very detailed knowledge. Was helpful, but unfortunately has no knowledge of English.

Constructor Captain S. MAKINO, of the Navy Technical Department, TOKYO, speaks English fairly well. Supplied information on layout of magazine and shell rooms and ring bulkhead construction.

5

Mr. MAENO, Engineer at Kure Ordnance Department before retirement to HIROSHIMA. Was responsible for the design of the cradle and slide of the 18-inch mount, but was of little value in giving information.

Lt. Comdr. MATSAMURA - Kure Ordnance Department; trained at Tokushima Technical College and was a specialist in gunnery engineering. Has fair knowledge of the 18-inch mount and speaks English reasonably well.

Harao KATAOKA - Assistant Engineer, Ordnance Department, Kure Naval Arsenal; supervised construction of guns.

Captain MITSUI, Head of Ordnance Experimental Department, KURE; ammunition and fuze specialist.

Captain Kichiro KURODA - Gunnery Officer of YAMATO at time of sinking. Had but a slight knowledge of the gunnery material of this ship.

6

INTRODUCTION

The answer to the long debated question, "What size guns did YAMATO and MUSASHI carry?" has been known for some time. They were 18-inch, 45 calibre guns. The objects of the investigation, which is the subject of this report, were:

1. To obtain data on the internal and external ballistics of the gun and general information on its method of construction.
2. To obtain as much detailed information as possible on the mount.

At a preliminary interview on 23 November 1945 with Captain IWASHIMA, IJN, Head of the Gun Section of the Navy Technical Department, TOKYO, a small amount of data on the gun was obtained. A list of some 50 questions on the salient features of the 18-inch guns and mounts and all others in use in the Japanese Navy since 1927 was presented to him. The answers to these were, unfortunately, not received until some five weeks later. In the meantime, investigations were carried out at KURE, where the guns and mounts were built. NavTechJap personnel at KURE had already collected a small pamphlet voluntarily prepared by Engineer Captain YASUNAMI, head of the Naval Ordnance Department of Kure Navy Yard. It gave a considerable amount of statistical data on the guns and mount, which after checking, and in many instances altering, have all been embodied in this report.

All documents and drawings of the guns and mounts are reputed either to have been destroyed by bombing or deliberatly burned. There has been no success in proving otherwise, so all information contained in this report is based on visual examination of the guns, and such part of the mounts as were available, and on discussions with Japanese personnel.

During the first interview with Captain YASUNAMI it was learned that a partially completed turntable and some 18-inch guns for SHINANO could be seen at one of the proving grounds at KAMEGAKUBI, and some turret training engines in the Navy Yard. This, to the best of his and his assistants' knowledge, was all that remained to be seen of these mounts. The first visit to the proving ground however, showed this to be incorrect, as another turntable and an upper and lower powder and shell handling room for a trial mount were also found. These also were only partially completed. During the subsequent visits, more and more pieces of turret machinery, packed in crates, were discovered. Most of this machinery was very inaccessible as these were in very heavy crates, stowed one above the other, and without a powerful crane (the one available was not in working condition) they could not be cleared away and opened up for examination. They were studied as carefully as could be done with the cases in place by breaking open the sides of the packing cases where possible.

Over a period of five weeks the Japanese at KURE were questioned on this mount and others, and put to work producing sketches of the more interesting parts of machinery.

Attempts to get the Japanese to produce a detailed and comprehensive description of the mount, or, for that matter, of any part of it, proved singularly ineffective, as only the most sketchy answers were produced. This was quite definitely not because they were trying to withhold information, but because they were incapable of thinking along those lines, and because all information had to be given entirely from memory.

7

Further information was obtained on returning to TOKYO in January 1946 from Engineer Captain DATE, a very capable engineer with a thorough knowledge of gun mounts, and a sound knowledge of the English language. Many ideas and impressions obtained at KURE were both clarified and corrected by him.

Part I of the report gives a general description and statistical data on the gun and mount, and Part II deals with the mount in more detail.

8

THE REPORT

Part I
HISTORY, GENERAL DESCRIPTION, AND DATA OF GUNS AND MOUNTS

A. HISTORICAL

1.   Shortly after World War I, Kure Naval Arsenal and the associated
Kamegakubi Proving Ground began work on a 48cm (18.9-inch)/45 calibre
gun. The gun was wire-wound only near the breech. The first gun split
during test firing, and a second was constructed; this was still at
KAMEGAKUBI in December 1945, (Figures 1 and 2), together with its cradle
and slide (Figure 3) which are copies of the British 15-inch design produced before World War I. Work on guns of this calibre was stopped by
the Washington Treaty and the subject remained in abeyance in Japan for
the next 12 years.

2.   In 1934 the topic of large naval guns was reopened in Japan with the
problem of the armament to be carried by the new battleships YAMATO and
MUSASHI. It was felt that 48cm was too heavy to be suitable, but that a
positive advance over 40cm (16-inch) would have to be made in order to
attain a definite superiority over United States (16-inch) gun battleships, and so a 46cm (18.11-inch) was selected. The gun (or, more
correctly, the breech mechanism) was designated Type 94 after the year
the design was commenced, and in order to keep its actual calibre a
secret, it was always referred to as 40cm/45 calibre Type 94.

3.   The design of the 18-inch guns and mounts was completed and produc-
tion commenced in 1939-40. Design work was largely the effort of the
late Engineer C. HADA, who was responsible for the design of most large
mounts for the IJN during the 40 years preceeding his death in 1943.

4.   It was originally intended to build three 18-inch battleships:
YAMATO, MUSASHI and SHINANO. The latter was altered to an aircraft
carrier while still building at YOKOSUKA. They carried nine 18-inch guns
in three triple turrets as shown in Figure 7. These ships had a standard
displacement of approximately 70,000 tons. Their designed speed was 27
knots with a shaft horsepower of 150,000, a waterline length of 256
meters (838 feet) and waterline beam of 36.9 meters (121 feet). YAMATO
attained a speed of 27.7 knots on trials.

5.   All 18-inch guns and mounts were built in the Ordnance Department of
Kure Naval Yard under the supervision of the late Engineer M. OYAMADA and
Adm. T. ITO. The number of mounts built is as follows: six complete
mounts, one trial mount (Figures 8 to 11) completed only sufficiently for
pit trials of turret machinery; one partially completed turntable for
SHINANO (Figures 54 to 57), and numerous portions of her turret machinery
including all her cradles and slides. (Figures 46 to 48, 51, 61).

6.   Some 27 of the guns were built in all. Eighteen were lost with
YAMATO and MUSASHI; two test guns at KAMEGAKUBI were demolished in
November 1945 in accordance with general disarmament orders of the United
States Army; and the remaining seven in various stages of completion were
found on the beach in a cove north of KAMEGAKUBI. Two of these have been
shipped to the United States, and the remainder were scheduled for demo-
lition. The seven had been under construction at Kure Arsenal for SHINANO
and the finished and unfinished guns and parts, as well as the special
tools for making them, were all moved to the cove and stored there with
very little maintenance or protection against corrosion.

9

Figure 1
19-INCH — 45 CALIBRE GUN
ON PROVING GROUND AT KAMEGAKUBI

Figure 2
(SAME AS FIGURE 2)

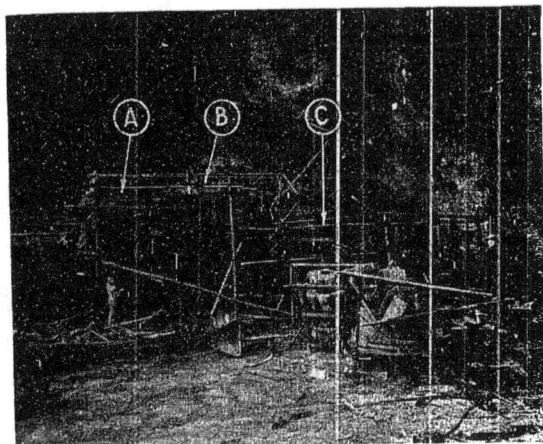

Figure 3
SLIDE FOR 19-INCH GUN
(A) Upper cap for rear cradle
(B) Front cradle
(C) Rear cradle (lower portion)

Figure 4
18-INCH GUNS AT KAMEGAKUBI
(BALANCE WEIGHTS NOT FITTED)

11

Figure 5
(SAME AS FIGURE 4)

Figure 6
18-INCH GUN (LESS BALANCE WEIGHT)
SHOWING ASSEMBLED BREECH MECHANISM

12

YAMATO

BOW

(1) ON TURRET

ON TURRET STERN

MUSASHI

BOW

(1) ON TURRET

ON TURRET STERN

NOTE:

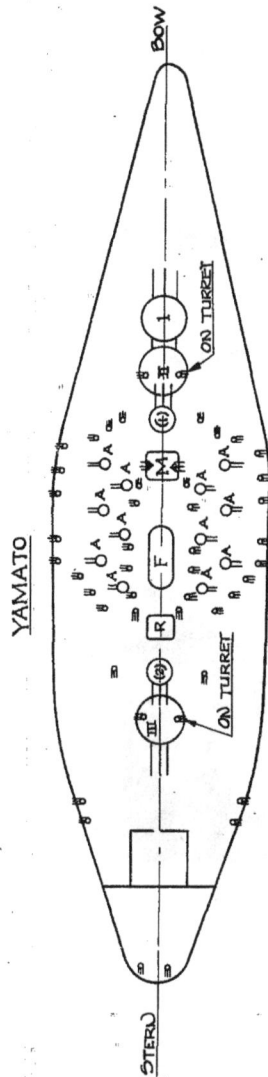

I, II, III	94 TYPE. 45 CAL. 46 CM. 8 TURRETS
(1) (2)	60 CAL. 15.5 CM. 8 TURRETS
AAAA	89 TYPE 40 CAL. 12.7 CM 8 AA GUNS
8	25 MM 8 MACHINE GUNS
M	MAIN MAST
F	FUNNEL
R	MIZZEN MAST
↓	13 MM 8 MACHINE GUNS

	YAMATO	MUSASHI
94 TYPE. 45 CAL. 46 CM. 8	3	3
60 CAL. 15.5 CM 8	2	2
89 TYPE 40 CAL. 8 127 CM A.A.	2	6
25 M.M. MG. 8	46	52
13 M.M. MG. 8	2	2

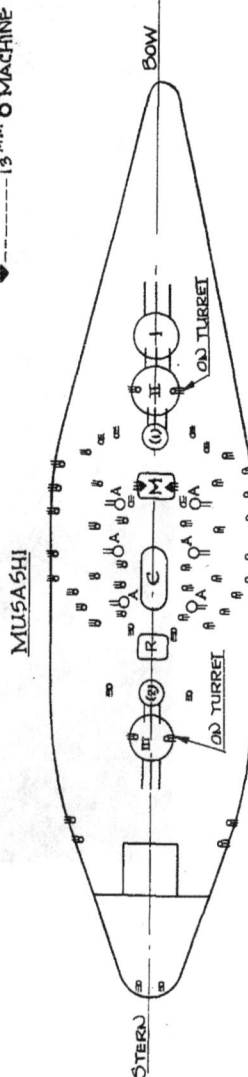

Figure 7

ARRANGEMENT OF TURRETS AND MOUNTS

13

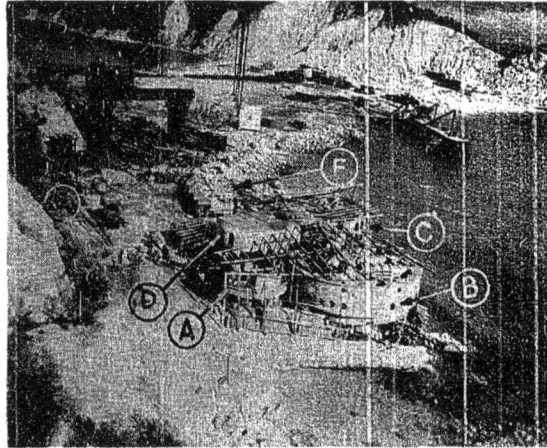

Figure 8
GENERAL VIEWS 18-INCH TRIAL MOUNT AT KAMEGAKUBI

Figure 9
(SAME AS FIGURE 8)

14

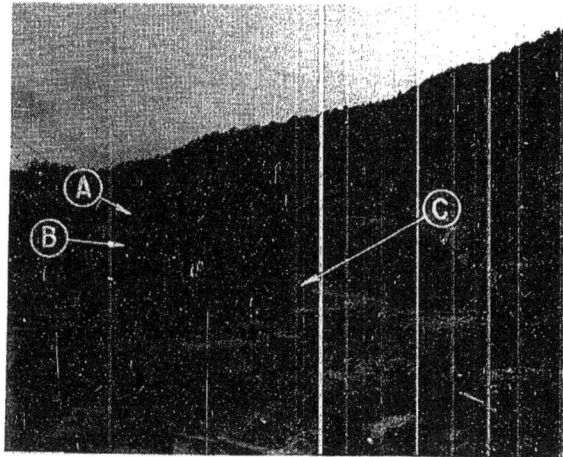

Figure 10
GENERAL EXTERIOR VIEW OF 18-INCH
UPPER AND LOWER SHELL HANDLING ROOMS (TRIAL MOUNT)
(A) Upper shell handling room
(B) Lower shell handling room
(C) Balance weights

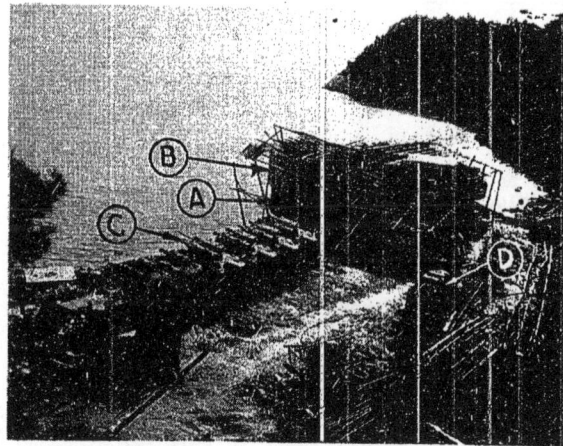

Figure 11
GENERAL VIEW SHOWING 18-INCH SLIDES AND ONE 16-INCH SLIDE
(B) Shell handling room
(C) 18-inch slides
(D) 16-inch slides

15

B.   GENERAL DESCRIPTION AND DATA

1.   Guns

a.   General:

Designation .... 40cm/45 calibre Type 94
Actual calibre 46cm (18.11-inch)
Length overall .... 69 feet 11½-inch
Weight, with breech .... 363,000 lb

b.   Construction: The gun barrels combined the processes of wire-winding and radial expansion. The #2A tube ran the full length of the gun; over this was shrunk the #3A, which extended about 3/5 the length from the breech, and the two were wire-wound with rings at the muzzle about half way back, the #4A, which covered about 2/3 the length from the breech, and the #5A, which covered the breech and the after part of the powder chamber. Belleville springs of silicon steel were fitted in the angles where the tubes changed diameter to take part of the stress caused by thermal expansion in firing. The #1A tube was finally radially expanded into place with hydraulic pressure of 110,000 to 120,000 psi. The bore was divided into three parts, and the breech third was given the greatest expansion.

2.   After inserting the liner, the bore was rifled with 72 grooves, 4.6mm (0.181-inch) deep, with uniform twist, one turn in 28 calibres.

When the liner was worn out, it could only be removed by machining out and inserting a new #1 tube. In practice this was so expensive a process that it was considered more practical to discard the gun without relining.

3.   Interior Ballistics:

Cross-section of bore .... 1698 sq cm (263.19 sq in)
Travel of projectile .... 17.590 m (57.71 ft)
Chamber volume .... 480 liters (29,290 cu in)
Service pressure .... 30 to 32 kg/sq mm (19.1 to 20.4 tons/sq in)

Weight of projectile
(AP and target) .... 1460 kg (3220-lb) 2m/d³...1.08
(IS and common) .... 1360 kg (3000-lb) 2m/d³...1.01
Muzzle velocity (3220-lb) .... 780 m/sec (2556 ft/sec)
(3000-lb) .... 805 m/sec (2640 ft/sec)
Service charge .... 330 kg (728 lb)

Reduced charges were also furnished for training and target rounds. They were designated "reduced" and "weak." "Reduced" gave about 2/3 of service velocity and was rated at 1/4 esr while "Weak" was rated at 1/2 esr. The life of the gun was not definitely determined, but was estimated at about 200 to 250 equivalent service rounds.

Service charges were cordite, made up in six sections per round. The bags used were wool until 1942, after which silk was adopted. Each bag had an ignition charge or 500 grams (1.1 lb) of black powder in a separate silk bag in one end, which was dyed red and was always loaded toward the breech.

4.   Exterior Ballistics:

No range table for the 46cm gun was recovered, and the Japanese officers interviewed claimed that all naval range tables were burned in August 1945. The following particulars have, however, been obtained for the trajectory of the 46cm gun with service charges:

16

Elevation	Range (yds)	Time of Flight (sec)
10°	18,410	26.05
20°	30,530	49.21
30°	39,180	70.27
40°	44,510	89.42
45°*	45,960	98.6
48°	46,050**	104
50°	45,790	106.66

* (maximum elevation of gun as installed)
** (maximum)

When mounted in a triple mount, with bore separation 350cm (11.48 feet) a delay of 0.08 seconds was introduced in the firing circuit of the center gun.

5. Mount and Miscellaneous Data

   a. Weights (In metric tons):

   Three guns and breech mechanisms .... 495 tons
        (363,000 lbs/gun, 165 tons/gun)
   Remainder of elevating parts .... 228 tons
   Turntable (less armor, guns, and elevating parts).. 350 tons
   Remainder of training parts below turntable .... 647 tons
   Gunhouse armor .... 790 tons
                                                       _____
        Total Revolving Weight   = 2510 tons

   b. Gunhouse armor thickness:

   Front 650mm (25.6-inch)
   Side 250mm (9.85-inch)
   Back 190mm (7.49-inch)
   Roof 270mm (10.63-inch)

   c. Dimensions:

   Diameter of roller path (Internal).. 11.500 meters (37.72 ft)
                           (External).. 13.050 meters (42.8 ft)
   Height from center line of guns to roller path.............
                                        4.40 meters (14.43 ft)
   Distance from center of rotation to trunnions.
        (In fore and aft line) .... 3.250 meters (10.68 ft)
   Distance between center lines of guns .... 3.50 m (11.48 ft)
   Length of recoil .... 1.430 m (4.69 ft)

   d. Rate of fire (See Part II, C, paragraph 27):

                    1.5 rounds per minute at maximum elevation.
                    Loading angle is fixed at 3° elevation.

   e. Power supply: Three turbo hydraulic pumps (and one standby pump) supply pressure at 1000-1100 psi through a ring main to center pivots below the lower cordite handling rooms. Water was used as a pressure medium. For further details of these pumps, reference should be made to NavTechJap Report, "Hydraulic Pumps in Japanese Naval Ordnance", Index No. O-53(N). No local pump was fitted in turrets for auxiliary purposes.

   f. Turret Rangefinder: 15 meter (49.2 ft), with an elevation of +10° and free training of 160 mils Right, 130 mils Left.

17

g. <u>Local sights</u>: 10cm telescopes (one per gun). Free training same as rangefinder.

h. <u>Ventilation</u>: Supply is by three 2.5 hp fans; exhaust by five 5 hp fans.

i. <u>Ammunition stowage</u>:(See Part II, C, paragraph 6):

    180 shells in revolving structure .... 60 rounds per gun
    120 shells in shell rooms .... 40 rounds per gun
    Total .... 100 rounds per gun

6. Figure 12 shows a sectional elevation of the turret, which is built and transported in five distinct sections, namely:

a. Center pivot and cable "lead in" gear (Figure 13)
b. Lower and upper cordite handling rooms (Figures 14 to 21)
c. Lower and upper shell handling rooms (Figures 10, 11, 27)
d. Turntable, complete with gunhouse deck and trunnion brackets. (Figures 8, 9, and 52 to 60)
e. Gunhouse armor.

7. <u>Magazines and Cordite Supply</u> (See Part II, B, paragraphs 1 to 14)

Magazines were fitted around both the upper and lower cordite handling rooms. The lower magazine supplies the center gun only, and the upper supplies the two outer guns by a cage and wire type hoist operated by a hydraulic winch mechanism. The cage was hoisted directly from handling room to gunhouse.

8. <u>Shell Rooms and Shell Supply</u> (See Part II, C, paragraphs 1 to 14)

Shell rooms and handling rooms were above the magazines. Half the quantity of shells carried were stowed in the revolving structure and half in the shell rooms, from which it was supplied to the lower shell handling rooms only. An original design of "push-pull" gear was used to move shells about in these compartments. "Pusher" type hoists, capable of being loaded in either of the shell handling rooms, raised the shells to the gunhouse. Shells were stowed vertically, being tilted horizontal only upon reaching the tilting bucket in the gunhouse. Separate gunhouse rammers were used for ramming shell and cordite, the latter being loaded in six one-sixth charges by a single rammer stroke.

9. <u>Turret Training</u> (See Part II, C, paragraphs 20 to 23)

Turrets were trained by special 500 hp vertical swashplate engines, through straight-toothed pinion gears, without the use of worm and worm wheel. Two entirely independent training motors were fitted in each turret, but only one was used at a time.

Maximum training speed = 2°/sec

10. <u>Elevating Gear</u> (See Part II, C, paragraphs 24 to 26)

Guns were elevated by a normal piston and cylinder type of elevating gear. The connection of the piston rod to the slide was, however, of novel design, in that it was movable longitudinally along the gun.

Maximum elevating speed = 6°/sec (designed) and 8°/sec actual
Maximum angle of elevating and depressing = + 45°, - 5°

18

11. <u>Recoil and Run-Out</u> (See Part II, C, paragraph 31)

Two recoil cylinders using glycerine and water, and two pneumatic run-out cylinders were fitted to each gun. An additional cylinder, also using glycerine and water, was utilized to control the speed of run-out.

12. <u>Ring Bulkhead and Armor</u>

The highest position at which the ring bulkhead was secured to the ship, was at the lower shell handling room level on the lower deck: from this point upwards, it was a free cylinder.

The main deck was the armored deck, being of 200mm (7.88 inch) armor. Magazines and handling rooms were protected from underwater explosion by a "triple bottom" of 75 to 80mm (3¼ inch) plating.

Part II
DETAILED DESCRIPTION OF MOUNT

A. <u>CENTER PIVOT, HIGH PRESSURE AIR, AND CABLE LEAD-IN ARRANGEMENTS</u>

1. Hydraulic pressure was led into the turret by a fairly normal design of center pivot. There were two pressure and two exhaust parts in the center fixed stalk of the pivot which was secured to the triple bottom (Figure 12). From the center stalk pressure was taken by a normal arrangement of parts to the external revolving portion attached to the lower cordite handling room and thence to the distributing panels in the lower shell handling room. The approximate overall dimensions of the center pivot were height 4 feet and diameter 30 inches.

2. High pressure blast air and electric cables were led in from the ship to a fixed inner tube (A-Figure 13). This tube (A) extended from the triple bottom, through the center pivot into the lower cordite handling room to within a short distance of the underside of the deck of the upper cordite handling room. At this point it was connected to a normal type of HP air center pivot and the HP air was led away into the revolving structure.

3. Electric cables came from the inside to the outside of the fixed tube (A) through four holes (B) just above the lower cordite handling room deck, and were then led into the revolving structure at the underside of the upper cordite handling room deck (C).

4. At the point (D), the external diameter of tube (A) was reduced to take the revolving sleeves E_1, E_2, E_3, E_4. After leaving the fixed tube (A), the cable (with adequate slack allowed for turret training) was secured to the bottom ends of sleeves E_1, E_2, E_3, E_4 by clamps F. As the turret was trained, the amount of twist in any one portion of the cable was limited to the desired amount by a slot (G) in fixed tube (A) and a tongue piece (H) on the bottom of the revolving sleeve E_1, and by similar slots and tongues on the remaining sleeves.

B. <u>MAGAZINE AND HANDLING ROOM CORDITE SUPPLY ARRANGEMENTS</u>

1. Figures 14 and 15 show a diagramatic arrangement of the magazines and handling rooms. Figure 16 is an enlarged view of a portion of the handling room.

2. When YAMATO was first commissioned, there were no watertight fore and aft bulkheads between the magazines as shown in Figures 14 and 15. It was not until after the Battle of the Philippines that they were fitted for damage control reasons.

19

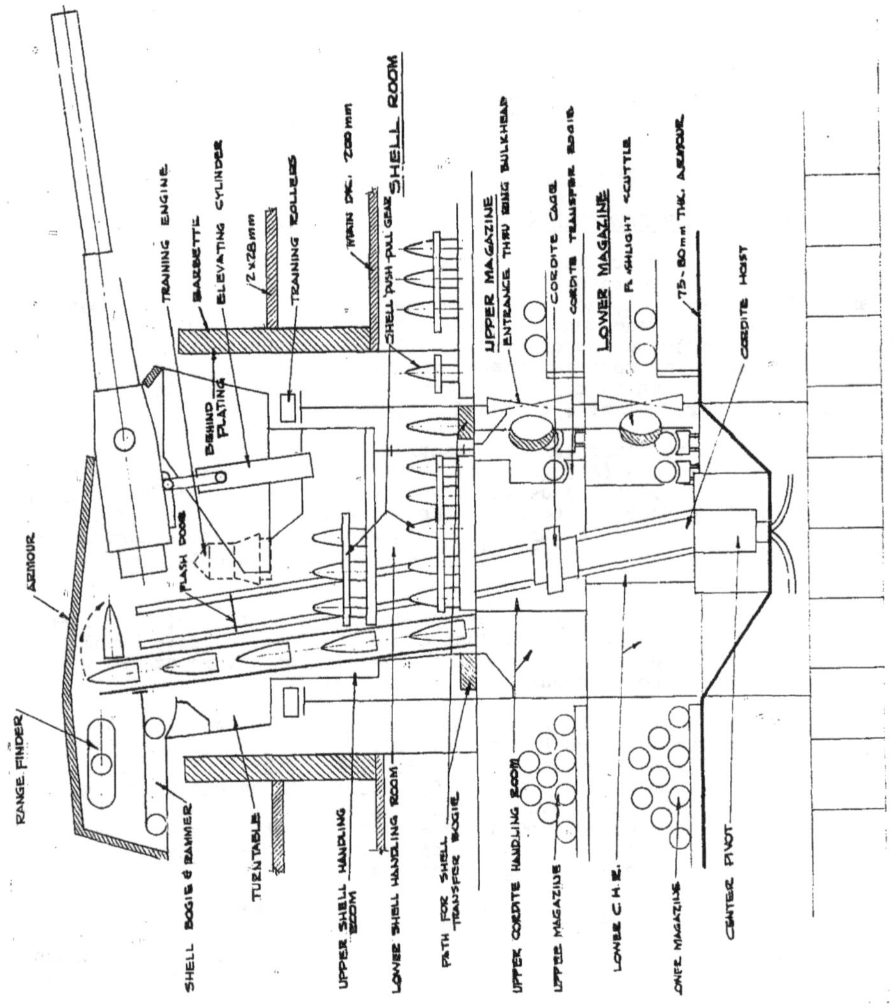

Figure 12

SIDE VIEW OF TYPE 94 – 18-INCH TRIPLE MOUNT

20

94 TYPE 45 CAL. 46 CM. GUN TURRET

CENTRAL PIVOT OF BLAST AIR

TRAINING PART

BLAST AIR PIPE

E₄

F

E₃

F

E₂

F

E₁

OUTER TUBE

INNER TUBE (FIXED)

F

H

G

D

ELECTRIC CABLE

BLAST-AIR PIPE

TURRETS CAN BE
TRAINED 360 DEGREES

D

D

B

ABOUT 3M

B   B

A

TRAINING PART

CENTRAL PIVOT OF PRESSURE & EXHAUST

FIXED PART OF SHIP.

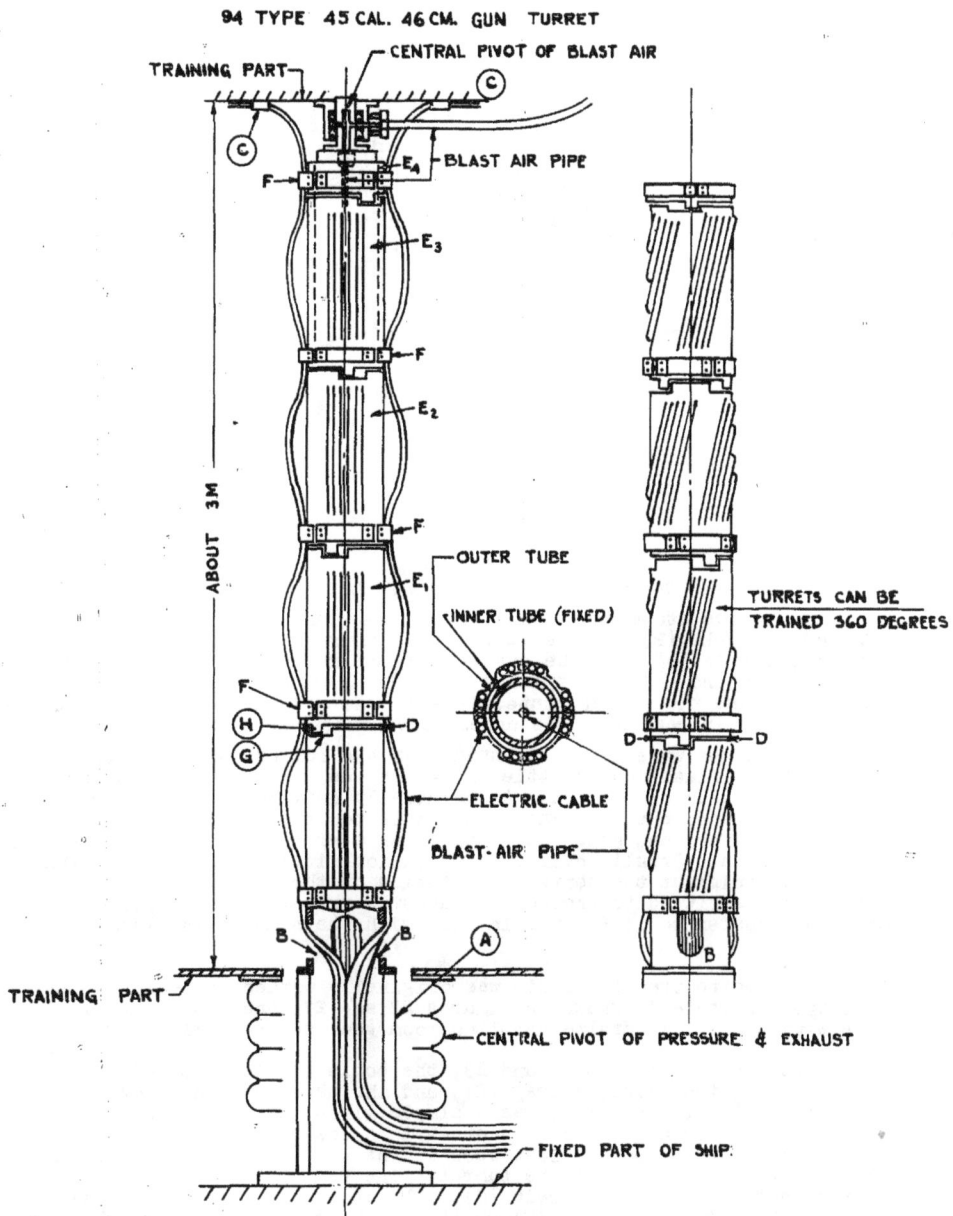

Figure 13
CABLE LEAD-IN GEAR

21

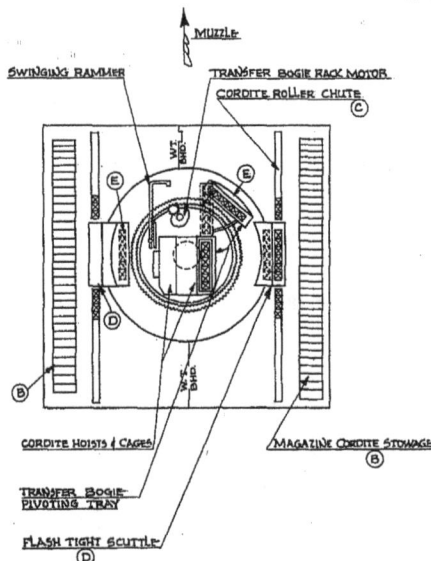

Figure 14

PLAN OF MAGAZINE AND UPPER CORDITE HANDLING ROOM

3. Cordite was stowed horizontally in one-sixth charges (weighing 121-132 lbs) in racks (B) (Figure 14). After removal from their canisters they were manhandled onto the cordite roller chute (C) on which they were loaded end to end. In the trial mount, the distance between the rack stowage (B) and the roller chute (C) was 60cms. This was found to be too small and was increased to distances varying between 80cm and one meter.

4. The charges were pushed by hand over the rollers of the chute into the revolving flashtight scuttle (D) which was long enough to hold six one-sixth charges end to end. As originally designed, the scuttle was revolved by hand, but this was found to be too slow, so a power mechanism was fitted. The hand gear was retained as a standby. The power mechanism consisted of a hydraulic cylinder and piston fitted with a rack, which engaged a pinion at the center of rotation of the scuttle. A mechanical interlock was fitted to prevent the scuttle from being revolved until the cordite transfer bogie (E) was in line with the scuttle and locked to the ship.

5. The cordite transfer bogie was originally worked entirely by hand. This type of bogie is shown in Figures 22 and 23, and was designed to work with the trial cordite handling room gear shown in Figures 17 to 21.

6. Referring to Figures 22 and 23, the bogie consisted briefly of the carriage (A), the pivoting tray (B), and the handwheel and traversing mechanism (C). The carriage was a simple girder structure mounted on wheels adequately angled to each other to allow the carriage to be moved on a circular path on the fixed deck of the cordite handling rooms around the revolving structure of the turret. The bogie was traversed by means of the handwheel and gear train (C). The lever (D) connected either one or the other of the pinions (E) or (F) to the gearing (C) through a simple dog clutch. One pinion (F) meshed with a circular rack on the fixed structure and was used for traversing the bogie from the flashtight

22

SECTION AT FR. 96

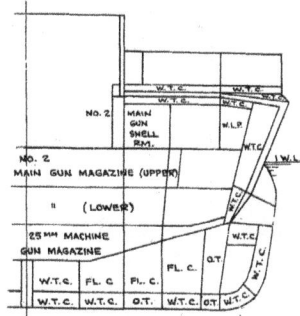

FLYING BR.

UPPER DK.
MIDDLE DK.

LOWER DK.

PLATFORM DK.

1ST HOLD DK.

2ND HOLD DK.

1ST HOLD DK.

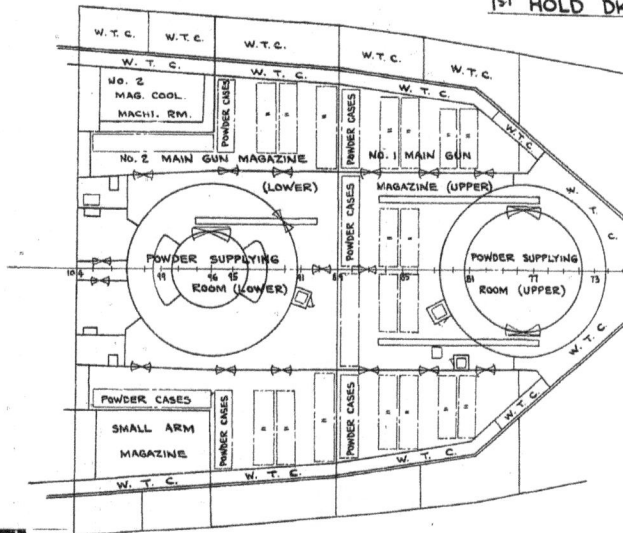

2ND HOLD DK.

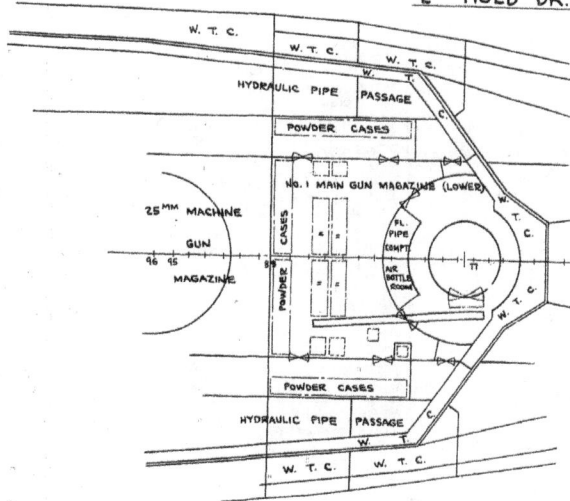

Figure

ARRANGEMENT OF MAGAZINES AND S
(No. 1 and No. 2

23

O-45(N)

PROFILE

NO. 2 TURRET

NO. 1 TURRET

STORE

NO. 2 MAIN GUN SHELL RM.

GYRO COMP. ROOM

W.L.

NO. 1 MAIN GUN SHELL RM.

W.T.C.

W.T.C.

BILGE PUMP ROOM

NO. 2 MAIN GUN MAGAZINE (UPPER)

NO. 1 MAIN GUN MAGAZINE (UPPER)

(LOWER)

(LOWER)

25 MM MACHINE GUN MAG.

OIL PUMP RM.

W.T.C.

W.T.C.

W.T.C.

O.T.

BILGE TANK

W.T.

BILGE TANK

105 103 101 99 97 95 93   87 85 83 81 79 77 75 73 71 69 67

LOWER DK.

W.T.C.

W.T.C.

W.T.C.

WIRE LEAD PASSAGE

FAN COMPT.

H.A.G. TRANSMITTING STATION

H.A.G. TRANS. STA.

MAIN GUN TRANS. STATION

SECONDARY GUN TRANS. STA.

SECOND. GUN TRANS. STA.

NO. 2 MAIN GUN

SHELL ROOM

MAIN GUN TRANS. STATION

SHELL SUPPLYING ROOM

GYRO COMP. ROOM

104   91   86 85   89 85   81   77   73

TEL. EXCHANGE ROOM

CELL ROOM

WIRE LEAD PASSAGE

PLATFORM DK.

W.T.C.

W.T.C.

W.T.C.

W.T.C.

W.T.C.

W.T.C.

NO. 1 MAGAZINE COOLING MACHI. ROOM

POWDER CASES

BLACK POWDER STORE

A.   S.

POWDER CASES

POWDER CASES

NO. 1 MAIN GUN SHELL RM.

W.T.C.

NO. 2 MAIN GUN MAGAZINE (UPPER)

P.C.

POWDER SUPPLYING ROOM (UPPER)

P.C.

P.C.

P.C.

SHELL SUPPLYING ROOM

W.T.C.

106   91   86 85   81   77   73

P.C.

P.C.

P.C.

W.T.C.

POWDER CASES

POWDER CASES

A.   S.

BLACK POWDER STORE

WIRELESS TEL. ROOM

W.T.C.

W.T.C.

W.T.C.

W.T.C.

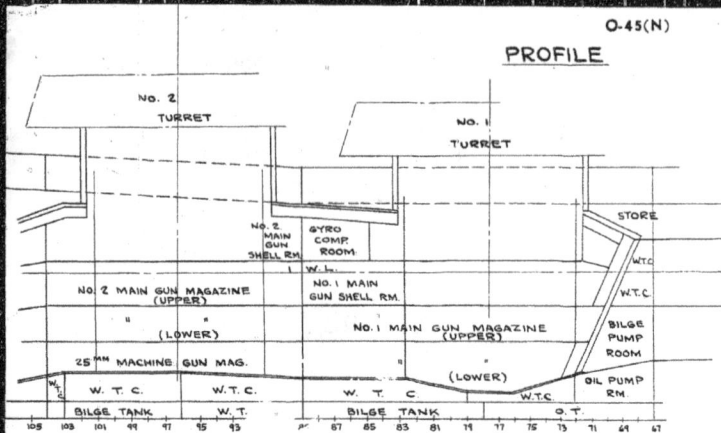

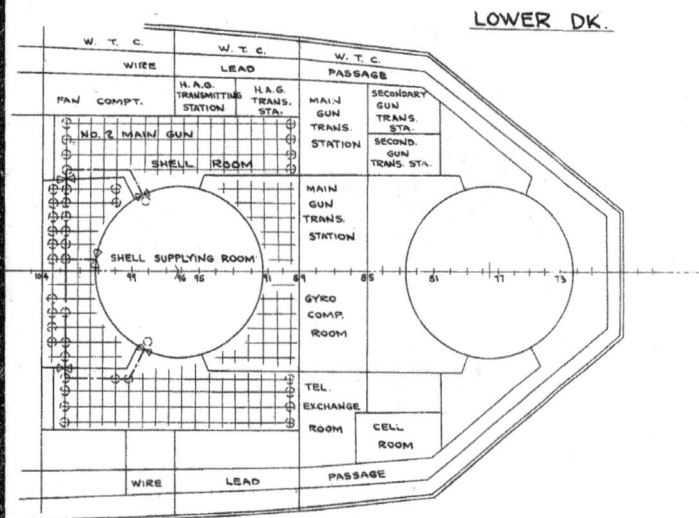

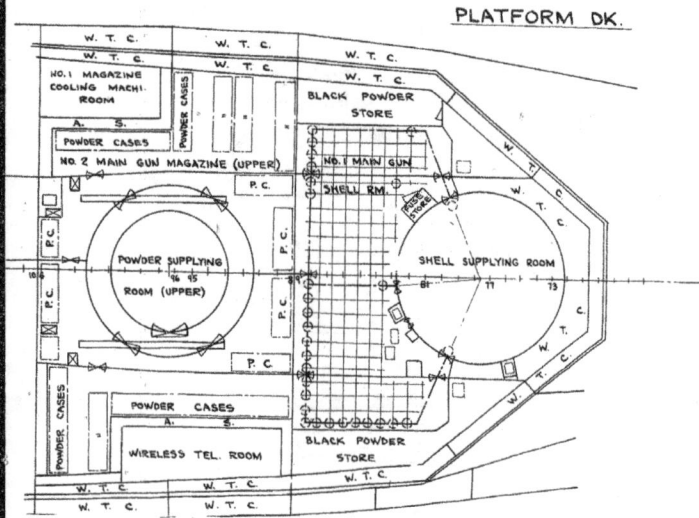

SHELL ROOMS – BB YAMATO
(Turrets)

BODY-26

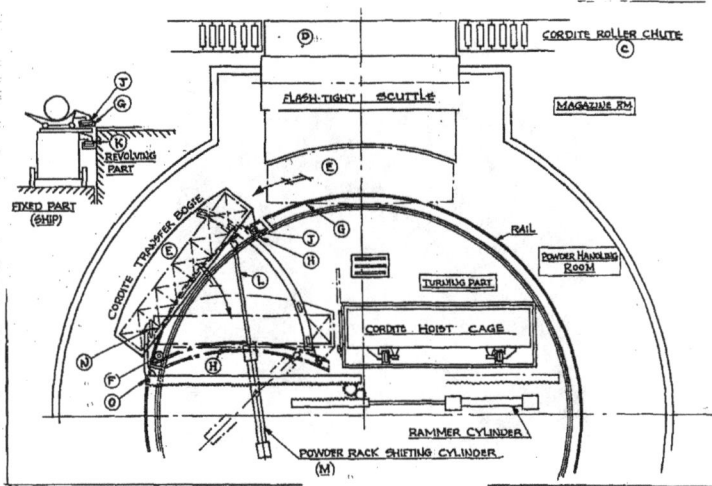

Figure 16
94 TYPE - 45 CALIBRE 46^{cm} GUN TURRET
POWDER HANDLING GEAR

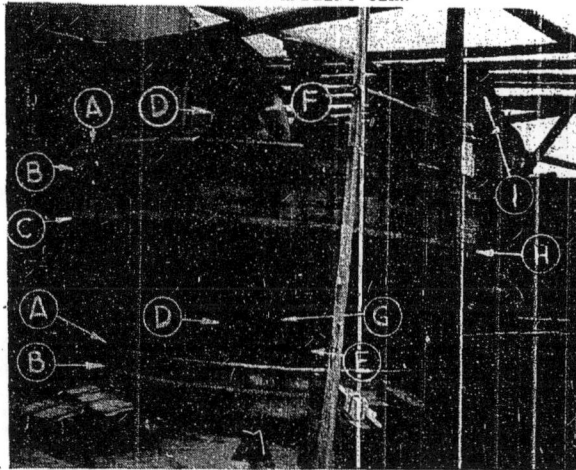

Figure 17
GENERAL VIEW OF UPPER AND LOWER CORDITE HANDLING ROOM
REVOLVING STRUCTURE OF TRIAL 18-INCH MOUNT
(A) Cordite bogie guide rails
(B) Cordite bogie traversing rack
(C) Roller path for vertical spring guide rollers
(D) Control levers for flash doors and cordite rammer
(E) Bed plates for swinging rammer
(F) Anti-flash door (right gun cordite hoist)
(G) Anti-flash door (entrance to center gun cordite hoist)
(H) Bottom of upper C. H. R. revolving structure
(I) Blast vent trunk

24

Figure 18
(SAME AS FIGURE 17)

Figure 19
UPPER CORDITE HANDLING ROOM OF TRIAL MOUNT

25

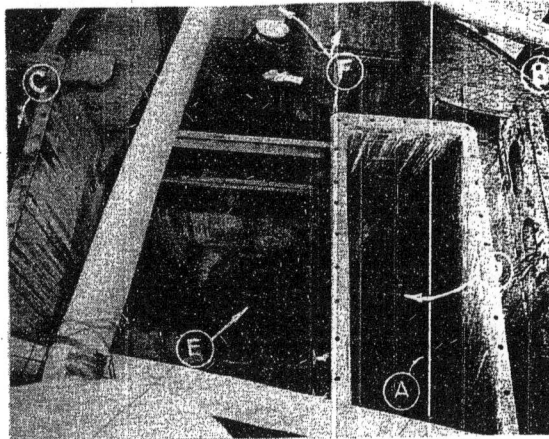

Figure 20

PLAN VIEW OF UPPER CORDITE HANDLING ROOM OF TRIAL MOUNT
(A) Center cordite hoist trunk showing flash door
    (D) at lower C. H. R. level
(B) Right cordite hoist trunk
(C) Left cordite hoist trunk
(D) See (A)
(E) Hole for securing center pivot
(F) Blast vent doors

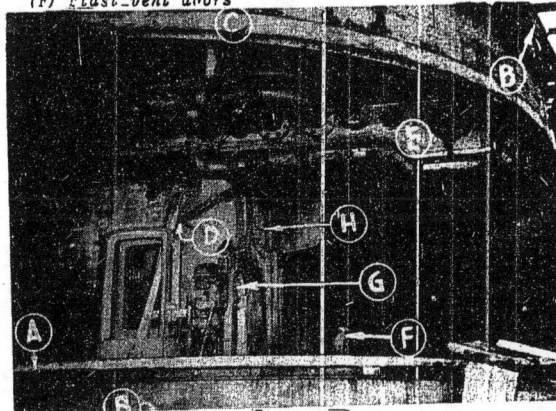

Figure 21

LOWER CORDITE HANDLING ROOM TRIAL DESIGN
(A) Guide vent for cordite traversing bogie
(B) Bogie traversing rack
(C) Roller path for vertical spring guide rollers
(D) Control levers for flash doors, cordite rammer
(E) Bed plates for rammer
(F) Rail for pivot tray roller
(G) Anti-flash door
(H) Inspection

26

scuttle to the position in line with the entrance to the hoist. A second circular rack fitted to the revolving structure meshed with the other pinion (E) and was used to make small adjustments to the position of the other bogie when it was being locked to the trunk. Numerous trials were carried out on this bogie to determine its suitability for use when the ship was in extreme positions of roll. These trials showed that a hand-powered bogie was impracticable with 10° list, and it was therefore never fitted.

7.   No samples or drawings were available of the powered transfer bogie designed after these trials and fitted in YAMATO and MUSASHI. The following description of its construction is based on a discussion with Captain DATE. In general appearance, it was not unlike the hand-worked bogie, except that it embodied a platform on which the operator was stationed with his controls. The power supply, a normal hydraulic motor, and its control valves were housed in the revolving structure of the handling rooms. The motor was connected to a single "driving" rack, capable of being revolved in either direction around the base of the revolving structure. A second similar but lighter "control" rack was connected to the motor control valve. The motion of this rack was limited to an amount sufficient to operate the control valve. Its position, and therefore that of the control valve, was regulated from the bogie by a lever and a differential gear, which was necessary in order to maintain the desired position of the control rack when the bogie was in motion. A third "fixed" rack was fitted to the fixed deck of the handling room, for final adjustment of the bogie's position when "locking to ship" for receiving ammunition from the magazine, and for traversing the bogie by hand in case of failure of the power gear. In the upper handling room where there were two bogies, five racks in all were fitted; namely, two driving, two control and one fixed rack. In order to avoid serious damage to bogies and racks due to bogies colliding with each other, trip levers were fitted which automatically declutched them from the driving racks on collision.

8.   On reaching the loading position, (Figure 16) the bogie came up against a buffer stop, and was declutched from the driving racks and clutched to the revolving structure. On arriving at this position, the pivoting tray guide roller (J) had run off the end of the fixed guide rail (G) onto the rail (H) which was pivoted about the point (F). Attached to (H) was the piston rod (L) of a pivoted hydraulic cylinder (M), the control position for which was approximately in the position occupied by the man in Figure 21. By means of this hydraulic cylinder, the rail (H) was swung about (F) to the position shown dotted in Figure 16, bringing with it the guide wheel (J). The pivoting tray on top of the bogie (E) was made to swing about its lower left hand corner (referring to position in sketch) and so brought the charges into line with the flash door of the cordite hoist ready for ramming into the gunloading cage.

9.   Anti-flash doors were fitted at the entrance to each hoist and were operated by one of the levers (D) shown in Figures 17 and 21. When the gun loading cage arrived at the bottom of the hoist (1) it operated through a cam and lever mechanism, and indicator in front of the rammer operator showing "cage down" or "cage up", and (2) removed a bolt locking the flash door control lever in the "closed" position. The lever was then put to "OPEN", which (1) locked, by rod gearing and pin, the cordite hoist control lever in the gunhouse in the "DOWN" position, (2) opened the flash door at the bottom of the hoist, and (3) freed the cordite rammer control lever.

10.   The cordite rammer, shown diagrammatically in Figure 6 is of the piston, rack and pinion type and was housed in the casing above the head of the rammer number on the left side of the hoist (not on the right as shown in Figure 16). The rammer head (N) was able to rotate in the ver-

27

Figure 22
CORDITE BOGIE (TRIAL DESIGN) IN HANDLING ROOM

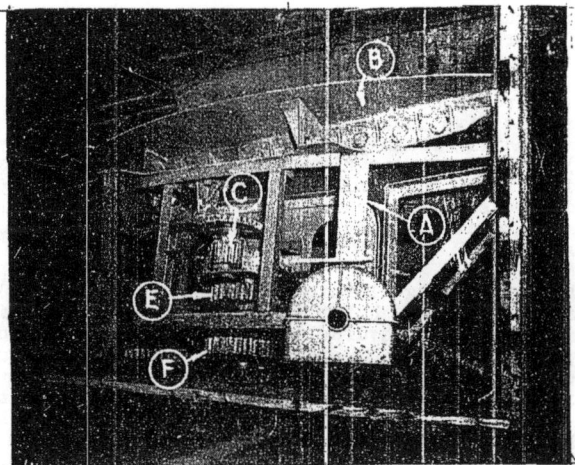

Figure 23
(SAME AS FIGURE 22)

28

tical plane about the point (O). It was normally in the "UP" position,
clear of the cordite transfer bogie. When the charge was ready for loading, the rammer head was swung downwards into line with the charge by a
crank operated by a roller working in a cam groove in the rammer casing.
After ramming the charge, the rammer was withdrawn, and the flashdoor
closed by the operator on the levers thus freeing the cordite hoist interlock in the gunhouse. The cage was then ready to be raised by the
operator in the gunhouse.

11. The lower and upper cordite handling rooms were similar except that
the lower had only one transfer bogie and one hoist (for the center gun)
while the upper had two bogies and two hoists (for the outer guns).
Around the base of the revolving structure in the upper handling room was
a machined path (C) in Figure 17 and Figure 21) on which ran the vertical
spring guide rollers for steadying the lower end of the turret revolving
structure. A blast vent trunk (Figures 17 and 20) was built into the
front of the upper handling room with light steel doors opening upwards
from the lower handling room for venting the force of any explosion in
either of these compartments into the lower shell handling room through
groups of holes in the shell handling room deck above the vent trunk.
The vent holes were directly below shells stowed in the front of the shell
handling room. From this handling room, the force of explosion was to be
vented upwards through shell hoist, manholes etc., and sideways through
holes and opening in the shell handling room walls into shell rooms.
This method of venting explosions in the cordite handling room was given
by Lt. Comdr. MATSUMURA and Mr. SUGIYAMA. When the writer expressed considerable surprise and doubt that the vent holes were directly below
shells stowed in the handling room they were adamant that this was so
(although the trial mount had not been so fitted). Captain DATE later
disagreed with their statement and said the vent trunks led into the
space between the fixed and revolving structures, a much more reasonable
explantion.

12. In Figure 20 a portion of the deck plating between the center (A)
and left cordite hoist (C) is removed, to give a view of the lower handling room and the hole (E) around which is secured the center pivot.

13. The cordite hoist consisted of a flashtight trunk three feet one inch
by nine feet five inches internal dimensions running straight from the
cordite handling rooms to the gunhouse (approximately 55 feet in "A"
turret). The hoist was fitted with guide rails to take a single cordite
cage (details of which will be found in paragraphs 15 to 18). An anti-
flash door was fitted in the hoist just above the turntable floor and was
operated automatically by the cordite cage on its passage up and down the
hoist.

14. The cordite cage lifting gear (Figures 24 to 26) was fitted to the
front of each cordite hoist in the lower shell handling room. The old
wire and sheave type hydraulic press had been discarded in favor of what
the Japanese consider a more efficient hoist. This consisted of a hydraulic cylinder and piston fitted with a rack extension, driving a train
of pinions connected to a wire-winch drum. An idea of the cumbersome
nature and weight of this gear can be gathered from Figures 24 and 26. A
straight-toothed rack (C), on the side of which were cam rails (A) for
the operation of the hoist control and cut-off valves, was connected directly to the cordite hoist piston. The rack driving through pinions (D'
(E), (F), (G) and (H) revolved the grooved drum (B) around which was
wound the cordite cage lifting wire. The wire was guided onto (B) by a
jockey pulley.

The overall velocity ratio of piston to cordite cage was of the order of
1 to 10.

29

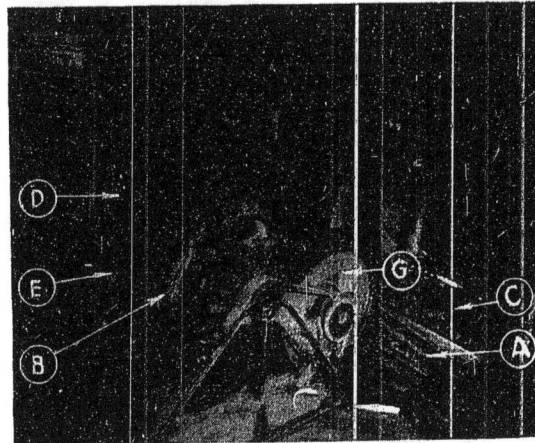

Figure 24
CORDITE HOIST RACK AND GEARING — BEFORE ASSEMBLY

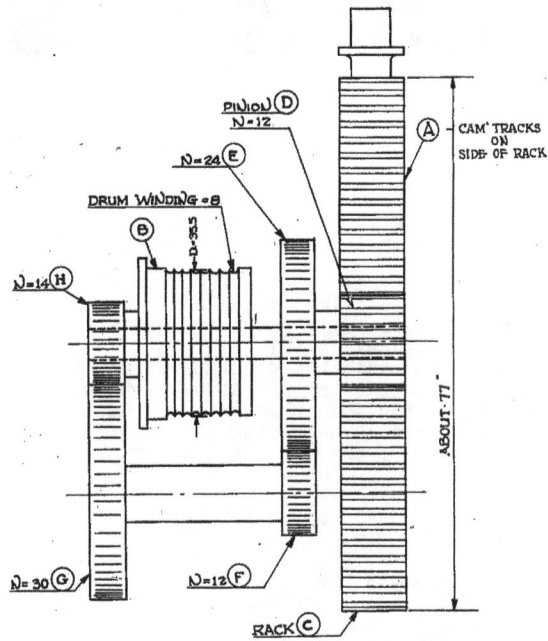

Figure 25
GUN TURRET POWDER HOISTING WINCH

30

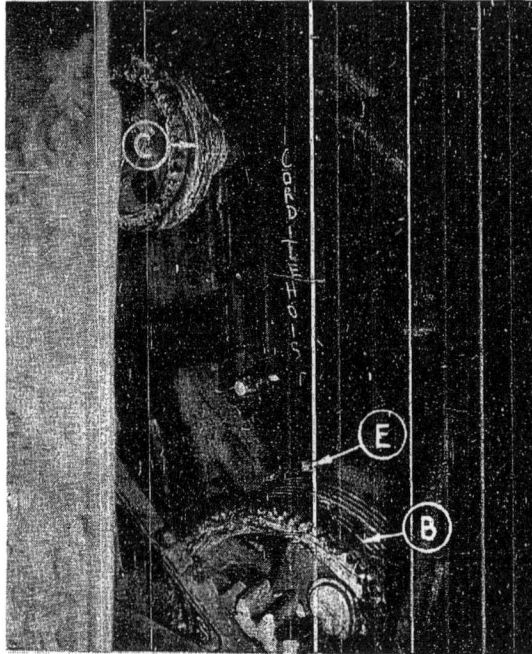

Figure 26
CORDITE HOIST WINCH

C.   SHELL SUPPLY

    1.   Shell Rooms and Shell Handling Rooms

The diagrammatic sketches in Figures 12, 14, 15, and 27 give some idea of the arrangements in the shell and shell handling rooms.

    2.   The main feature of interest, is the unusual method used for moving shells from the stowage position to the hoists. The gear used for this purpose was known as "Push Pull" gear. (Figures 28 to 31). Shells were stowed vertically on twin skids (C) (Figures 28 and 29) and between heavy girders (G). The whole available deck space in shell rooms and shell handling rooms, was occupied by this stowage which was arranged either in the fore and aft direction, or athwartships. Suitable junctions (X), Figure 31, or breaks in the run of the girders (G) allowed the direction of movement of shell to be changed from athwartships to fore and aft, or vice versa.

    3.   The shells were held and moved between the girders (G) by two "gates", (A) and (B), upper and lower.

(A), the upper was known as the "fixed gate", and was capable only of rotation in the vertical plane, through an angle of 90° about the shaft to which it was keyed. (See Figure 31) When (A) was in the "down" or horizontal position the fingers of the gate were around the shell, just below the ballistic cap. The function of the gate (A) was simply to support the shells against rolling and pitching and to keep them at the correct

31

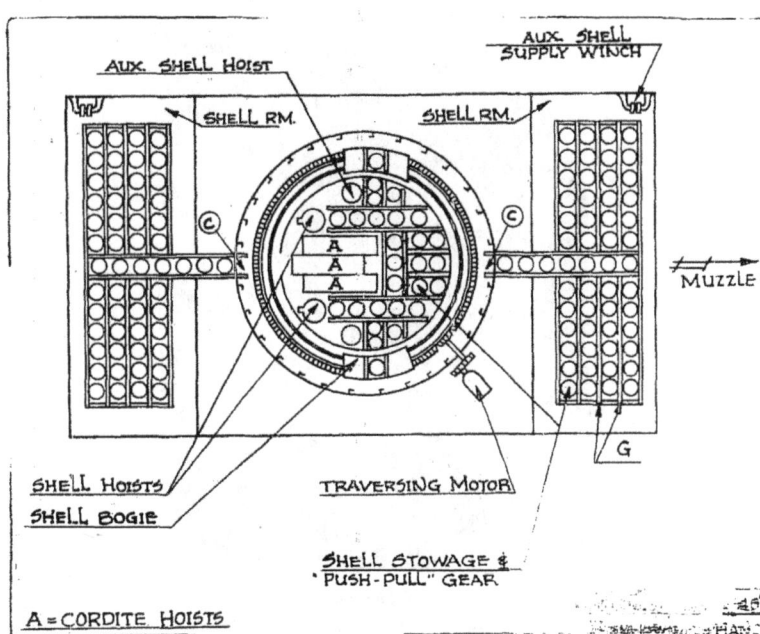

Figure 27
LOWER SHELL HANDLING ROOM – DIAGRAMMATIC PLAN

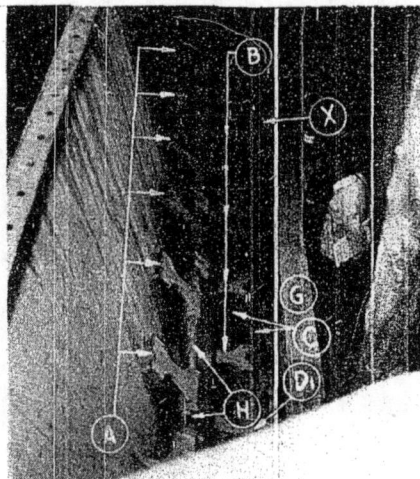

Figure 28
SHELL "PUSH PULL" GEAR AND SHELL HANDLING ROOM

32

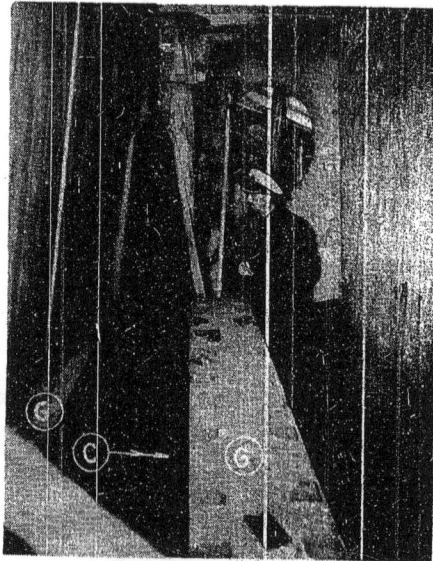

Figure 29
ENTRANCE TO SHELL HOIST FOR CENTER
GUN IN UPPER CORDITE HANDLING ROOM
("Push-Pull" gear not fitted)

Figure 30
UNIT OF SHELL "PUSH PULL" GEAR (TRIAL DESIGN)
FOR HANDLING ROOM

33

distance from each other when not being supported by the lower gate (B) which was known as the "moving gate" and was capable both of rotation through 90° in the vertical plane about a shaft level with the center of gravity of the shell and of moving bodily backwards and forwards between the girders (G). In the "secured" position, fixed gate (A) was down, and moving gate (B) was up completely clear of the shell. When moving a shell, the action of the gear was as follows:

    a. Moving gate lowered.
    b. Fixed gate raised clear of shell.
    c. Moving gate moved horizontally one shell space, taking all
        shells in the gate with it (six in the gate shown in
        Figure 28).
    d. Fixed gate lowered.
    e. Moving gate raised.
    f. Moving gate returned one shell space ready for the next cycle.

In the trial mount the gates were connected by 90° link gearing so that they rotated as one unit, 90° out of phase with each other. This was found to be unsatisfactory, and in the final design they were completely disconnected and each given its own control lever and hydraulic operating cylinder. In the final design the size of the gate fingers was considerably reduced.

Figures 28 and 31 show the hooks (D) and (D_1) at each end of a gate. The function of the hooks (D) was for one to pull a shell out of an atwartships bay into a fore and aft bay and of the (D_1) to push a shell from the end of its bay into the shell hoist. Hence the name "Push-Pull" gear.

4.   The control mechanism for the "Push-Pull" is shown in Figure 30. It consisted of normal hydraulic cylinders and pistons (E) and (F) rotating the gates through suitable bell cranks, and moving the gates horizontally, either by rack and pinion or by a simple tug on the end of the piston rod that hooked into a slot (elongated radially to allow for 90° rotation) in the connecting bar of the moving gate.

5.   <u>Embarking Shell</u>

The shell was embarked into the shell room by a conventional grab capable of picking up a shell horizontally on the upper deck, swinging vertically, and lowering and disengaging in this position on the skids of the shell room deck. By use of "Push-Pull" all bays in the shell rooms could be filled and shells moved step by step up to the positions (C) (Figure 27). In the fixed structure around the outside of the lower shell handling room there were two shell transfer bogies (Figures 27 and 32). The shell was loaded into one of these bogies by the "Push-Pull" at point (C) and traversed either by power or by hand to one of the three entrances to the shell handling room, (A) in Figure 10 and 11. In this position the bogie was locked to the turret and the shell pulled out of it by the end hook of a "Push-Pull" unit in the handling room. In this way, all positions in the lower shell handling room might be filled. The only way of getting a shell into the upper shell handling room was by loading first into the lower handling room and then lifting by means of the auxiliary shell hoist or the main hoists into the upper handling room, a very slow and laborious method of ammunitioning ship. Two estimates were obtained of the time taken to ammunition ship, but they were somewhat unreliable. The first was given by Lt. Comdr. MATSUMURA at KURE, who said the ship could be completely ammunitioned in 24 hours, working day and night. The second was by Captain K. KURODA, Gunnery Officer of YAMATO at the time of her sinking. He said four days were required, working during daylight hours only. He had been gunnery officer of YAMATO for only three weeks, had never taken on ammunition and seemed otherwise poorly acquainted with her armament, but this seems to be the better estimate. Cordite was

34

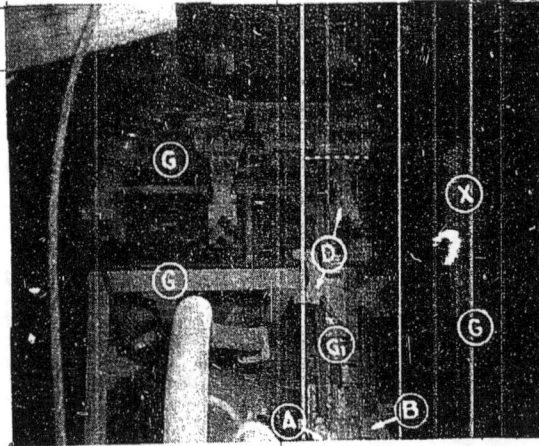

Figure 31
18-INCH MOUNT – SHELL "PUSH PULL" GEAR (TRIAL DESIGN)

Figure 32
SHELL TRANSFER BOGIE

35

loaded via normal embarking hoists directly to the magazines.

6.   Shell Stowage

The requirements given when designing the ship were to provide stowage
for a total of 100 rounds per gun. It was intended that enough rounds be
stowed in the revolving structure to fight a surface action, without having to transfer shells from the shell rooms to the shell handling rooms,
since, with the gear as fitted, this would have been a very slow procedure. Arrangements were therefore made to stow 60 rounds per gun (total
180 rounds) in the revolving structure and 40 rounds per gun in shell
rooms in addition to a certain number of practice shells. This involved
stowing 120 rounds in the lower shell handling room, outer shell hoists
and outer gunhouse waiting positions, and 60 rounds in the upper shell
handling room, center hoist and waiting positions. As the handling rooms
inspected at KAMEGAKUBI were built for trial purposes only and were
therefore only partially completed, it was not possible to verify that
these numbers of shells could, in fact, be stowed.

7.   Shell Hoists

The shell hoists were simple "pusher" type hoists fitted with a single
set of moving, or lifting, pawls in back and fixed retaining pawls at the
front. There was no scuttle at the bottom of the hoist. Shells were
pushed by the "Push-Pull" directly into the hoist through a suitably
shaped opening (Figure 29), and toppled onto a platform sloped at about
5° below the horizontal, to bring them into line with the hoist which is
raked at 5° from the vertical. A spring-loaded crank, projecting above
the platform, was depressed by the shell on entering the hoist. This
crank was connected to anti-rolling stops at the sides of the hoist entrance (not fitted in hoist in Figure 29) which were thus closed, to prevent the shell falling out of the hoist. A slot was cut in the platform
to allow passage for the lowest lifting pawl. All three hoists were
carried to the bottom of the lower shell handling room. There was an
entrance to each hoist both in the upper, and in the lower shell handling
rooms. Shells were normally supplied from the lower S.H.R. and during
action more shells were to be supplied from the shell rooms. The rate
of supply from the shell rooms was not normally sufficient to keep up
with the rate of supply from the handling rooms to the gunhouse. When
the supply from the lower handling room was exhausted, it was continued
from upper S.H.R. by inserting portable platforms in the hoist at the
entrance at that deck level.

8.   The piston of the lifting cylinder was fitted with a rack extension,
which was connected through a 2 to 1 ratio pinion gear to a rack fitted
to the lifting pawls' connecting lead. The stroke of the piston was approximately 1.3 meters (43 feet) and of the pawls 2.6 meters (85 feet).
The moving pawls, of which there were four, were all positively operated
by a roller running around an "island" cam on the inside of the hoist
trunk. The "fixed" pawls were all interconnected and were pushed into or
pulled out of the hoist by bell crank and rod gearing operated by a roll
running in a cam groove on the side of the rack extension of the piston rod. There were three fixed pawls in each hoist.

9.   Shell Hoist Interlock

The only interlock fitted was manually operated by a man standing near
the entrance to the hoist. He had a lever which was mechanically connected
to tell-tales and locking pins in (1) the "Push-Pull" operators position, where it showed whether or not the hoist was clear for ramming so
that his lever could be locked if it was not clear, and (2) in the gunhouse to prevent the hoist being raised before shell was loaded into the
hoist.

36

10. Gunhouse Shell Tilting Bucket and Waiting Tray (Figures 33 and 34.)

The top pawl of the shell pusher hoist, lifted the shell into a "U"-shaped
tilting bucket well to the side of and above the center line of the gun
bore when at the loading angle. A spring operated pawl returned the
shell in the bucket when the pusher hoist started its down stroke. There
was no interlock to prevent the shell hoist being raised when the tilting
bucket was in the "down" position. Good drill was relied upon to prevent
such an accident. This was not considered to be a serious deficiency
when base-fused shells were used, but with nose-fused shell a safety interlock was thought to be very necessary; one was being designed. It was
never fitted.

11. The bucket was tilted down to 8° above the horizontal by a hydraulically operated rack, crank, and connecting rod as shown in Figure 34. The
reason the bucket was not tilted down to 3° (the loading angle of the
gun) will be explained later in the discussion of the shell bogie and
rammer. The shell was prevented from falling out of the side of the bucket by retaining clips (A) in Figure 34. These clips were opened by link
gearing and a roller working in the cam rail (B) on the side of the waiting tray. This allowed the shell to roll out of the bucket onto the
waiting tray, where it was held by the stops (C) until it was rolled onto
the receiving tray of the combined shell loading bogie and rammer.

12. Shell Loading Bogie and Rammer (Figures 35 and 36.)

The shell loading bogie and rammer was a large unwieldy contraption weighing about three tons and bearing a marked resemblance to some prehistoric
monster (Figure 35). Its main dimensions are shown in Figure 36. It was
mounted on four wheels running on rails parallel to the bore of the gun.
In the stowed position, the upper surface of the bogie and the receiving
tray (A) for the shell were inclined at 8° above the horizontal. As the
bogie was moved forward toward the gun the front wheels ran down a ramp
to the position (B), thus bringing the bogie and shell to an angle of 3°
above horizontal, the loading angle of the gun. This arrangement was
necessary to allow room for the cordite rammer and gunloading cage cylinder to swing into line with the bore underneath the receiving tray (A)
when the shell bogie was in the stowed position. A breech thread pro-
tecting tray (C) was fitted on the bolt (E). In Figure 35 the tray is
not in position. Power for moving the bogie and for ramming was supplied
through telescopic pipes along the sides of the bogie. A buffer stop (D)
limited the forward movement of the bogie. As the buffer was pushed in
and met the breech face, anti-rolling grips (F) were opened to free the
shell preparatory to ramming. The total forward movement of the bogie
was about 13 feet, and it was kept in the forward position while ramming
by hydraulic pressure in the bogie moving cylinder.

13. The rammer was a normal chain-type rammer driven through a spring
shock absorbing mechanism (G) by pinion, rack, and hydraulic piston. Hydraulic buffer stops were fitted at each end of the rammer. A rigid bar
(H) extended for part of the rammer stroke to support the chain.

14. Figures 37 and 38 show the shell bogie and rammer as originally designed. This bogie, had three pairs of wheels. The main body (A) containing the rammer and its operating mechanism was mounted on the two
rear pairs of wheels, while the receiving tray and remainder of bogie (B)
was supported by the front pair. As the bogie was pushed forward, the
front wheels ran down a ramp allowing the receiving tray (B) to pivot
about the point (C), thus bringing the receiving tray and shell down into
line with the rammer head at the loading angle. This bogie was fitted
with a friction drive for the rammer instead of the spring shock absorber
on the later type. Both types of bogie have been sent to the United
States for further examination.

37

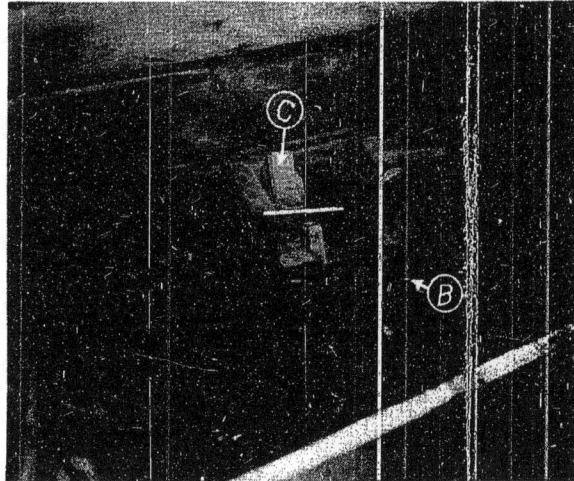

Figure 33
GUNHOUSE SHELL WAITING TRAY

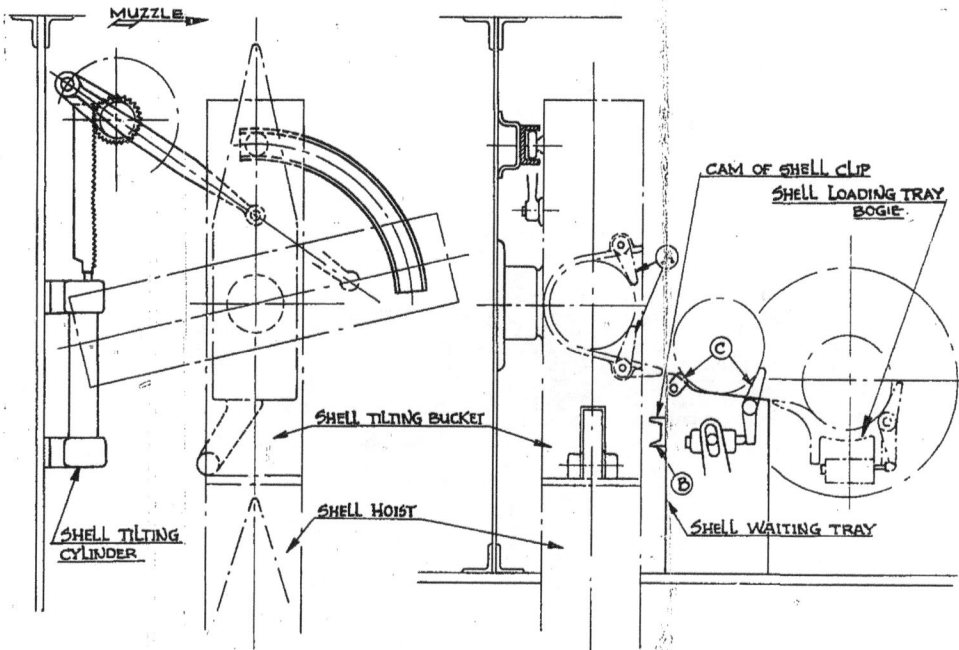

Figure 34
GUN TURRET SHELL TRANSFER IN GUNHOUSE

38

Figure 35
GUNHOUSE SHELL BOGIE AND SHELL RAMMER.- FINAL DESIGN

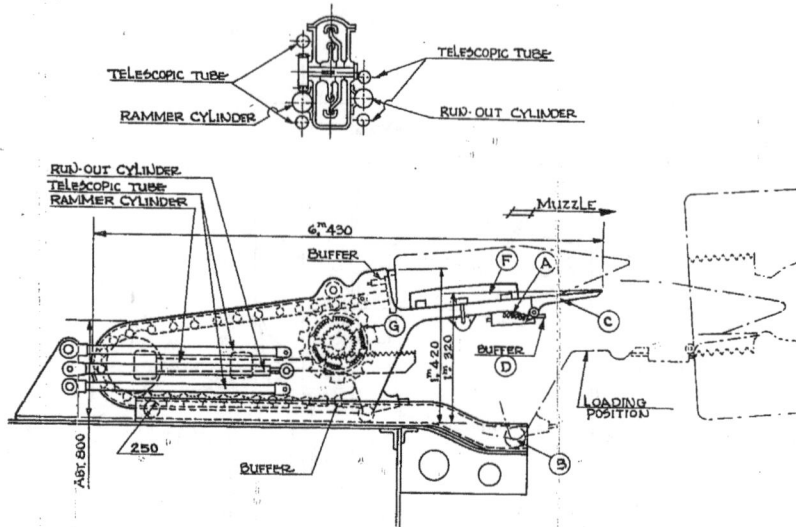

Figure 36
GUN TURRET SHELL RAMMER IN GUN-HOUSE

39

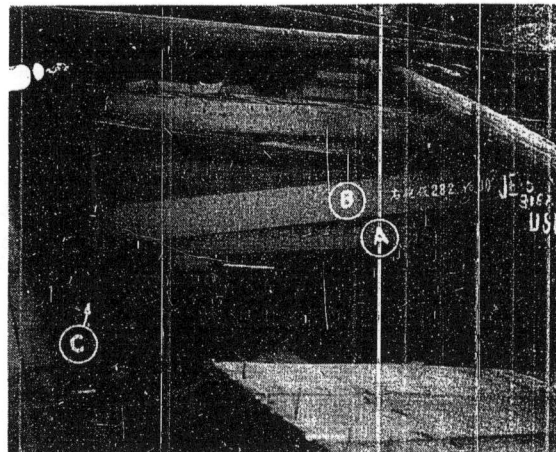

Figure 37
GUNHOUSE SHELL POGIE AND RAMMER (TRIAL DESIGN)

Figure 38
(SAME AS FIGURE 37)

40

15. Cordite Cage and Rammer (Figures 39 and 40.)

The method of getting the charge from the magazine into the cordite cage has been described in previous paragraphs. The description of the cage itself and the means of loading the charge into the gun has been omitted until after the description of shell loading in order to maintain the correct sequence of loading operations.

16. A cylindrical flashtight charge container (A) designed to hold six one-sixth charges in line was carried in the upper part of an open framework (B), which was mounted on wheels (C) and (K) running in suitable guide rails in the hoist trunk. The container (A) was mounted on wheels (G) running in rails formed in the top of a carriage (H), which was keyed to the shaft (D) and coupled to bearings at the bottom of the frame (B). The shaft (D) was offset from the center line of container (A), whose weight, with that of the charge, tended to keep them in place in the frame (B) on the way up and down the hoist. A lever (E), keyed to the shaft (D), and roller (F), running on a rail (J), also assisted in this (though this was not their main function).

17. When the cage approached the top of the hoist, roller (F) entered a portion of the rail (J) which was pivoted about and keyed to the shaft and crank (L), rotated by the connecting rod, pinion, rack and hydraulic cylinder mechanism (M). In order to bring the containers into line with the bore of the gun, the rail (J) was rotated to the position shown dotted in end view (Figure 39 (A)); (J) took the roller (F) with it, thus rotating crank (E), shaft (D), and carriage (H) to a position outside the main frame (B). During the last part of this rotation, a forward motion (relative to the carriage (H))was imparted to the container (A) by the pivoted link bars (Q) and (Q1) and link (R). This motion was controlled by a roller (N) on link (Q) working in cam (P). (A) was thus taken sufficiently far into the gun to cover the breech threads.

18. The container was made flashtight by end doors fitted to the main frame; these were opened and closed sufficiently by the rod of lever gearing (S) to allow free sideways movement of the container. A complete cordite cage has been sent to the United States for examination.

19. Cordite Rammer

(Figures 39 and 40) was fitted to the rear of and on the opposite side of the gunwell to the cordite cage. It was made to swing about its lower edge in order to bring the rammer head on line with the bore. The position of the rammer body was controlled as shown in Figure 39 by the same hydraulic cylinder (M) which controlled the swinging of the cordite container. In all other respects, the rammer was a conventional chain-type rammer worked by hydraulic piston and rack. In the trial design a telescopic piston-type rammer was used, but this was found to be too slow (8 secs per stroke) and the chain type with a speed of three seconds per stroke was substituted.

20. Training Gear (Figures 41 to 45.)

Two entirely independent sets of training gear were fitted 180° apart in each turret but only one was used at a time. In the opinion of the Japanese there would be two main disadvantages on designing a normal type of worm and worm wheel driven training gear for a mount of this size. First, it would require too much horizontal space, necessitating a very large diameter turret and, secondly, the load on the worm would be very high and it was feared that severe pitting would take place at the point of contact. It was therefore decided to design a wormless training drive. A diagrammatic sketch of this is shown in Figure 41. Details of the training pinion and rack are given in Figures 42 and 43. The details in Figure

41

42

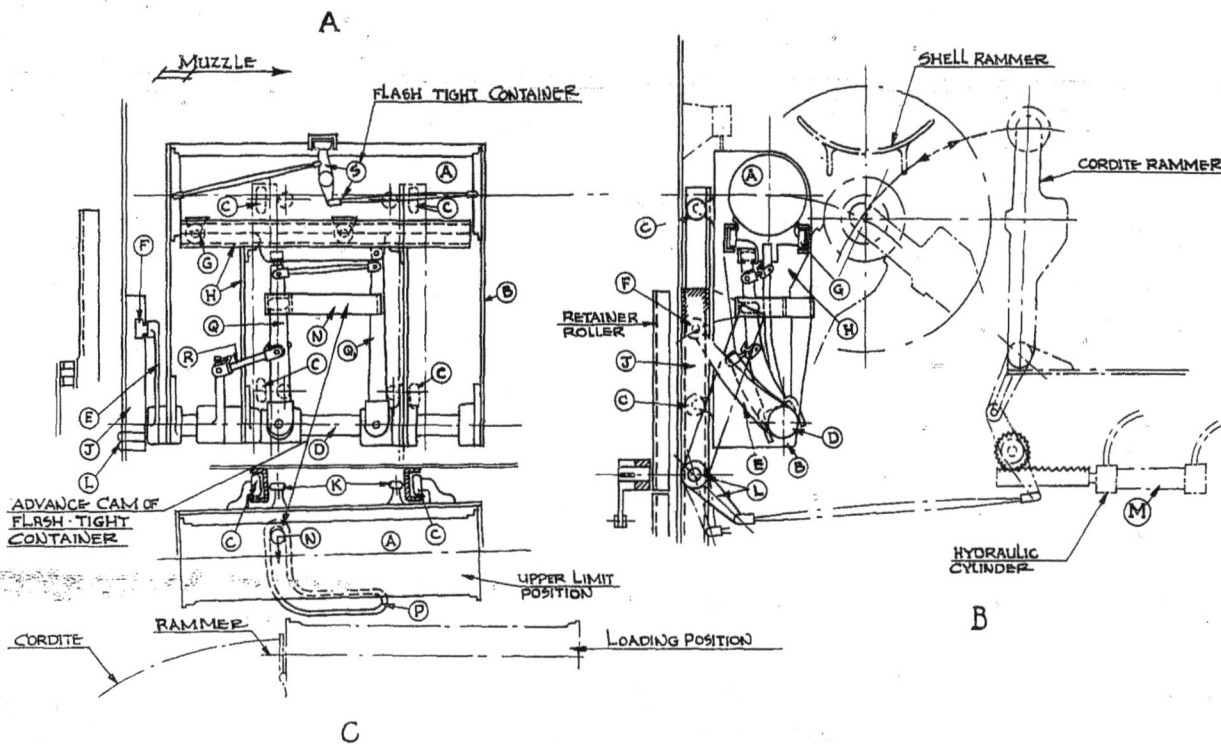

A

MUZZLE

FLASH TIGHT CONTAINER

ADVANCE CAM OF
FLASH·TIGHT
CONTAINER

CORDITE

RAMMER

C

UPPER LIMIT
POSITION

LOADING POSITION

SHELL RAMMER

CORDITE RAMMER

RETAINER
ROLLER

HYDRAULIC
CYLINDER

B

Figure 39
94 TYPE - 45 CALIBRE 46^{cm} GUN TURRET
CORDITE CAGE

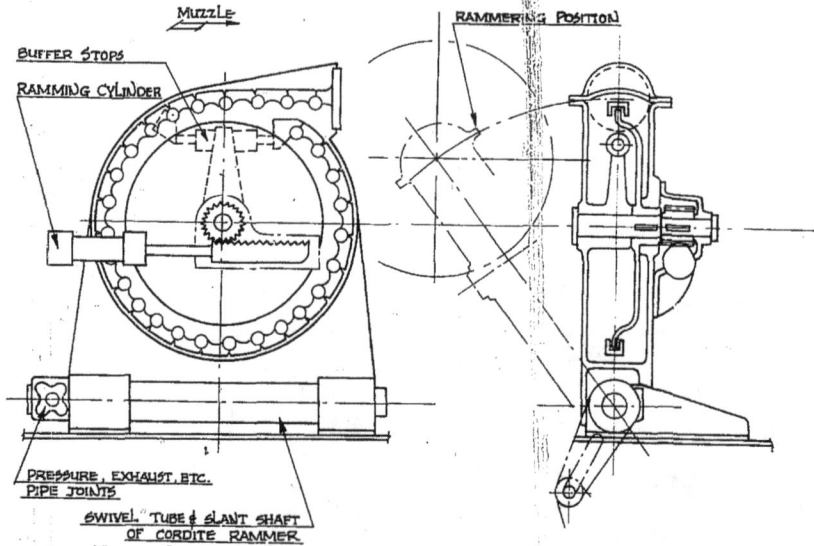

Figure 40
GUN TURRET CORDITE RAMMER IN GUN-HOUSE

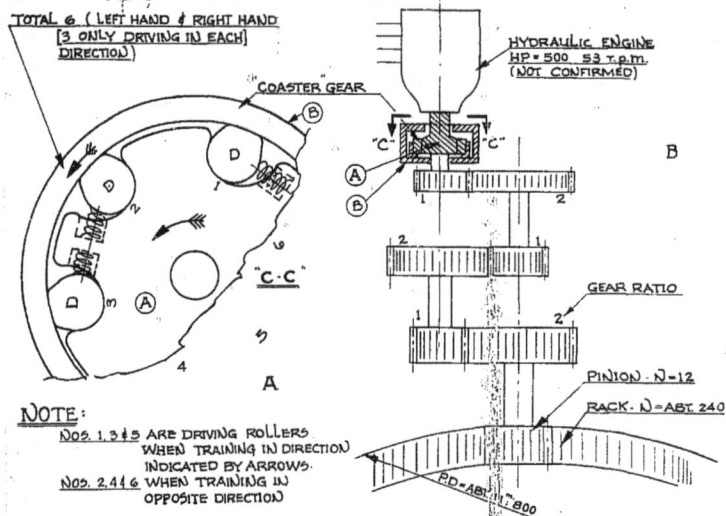

NOTE:
  NOS. 1, 3 & 5 ARE DRIVING ROLLERS
    WHEN TRAINING IN DIRECTION
    INDICATED BY ARROWS.
  NOS. 2, 4 & 6 WHEN TRAINING IN
    OPPOSITE DIRECTION

Figure 41
GUN TURRET TRAINING GEAR

43

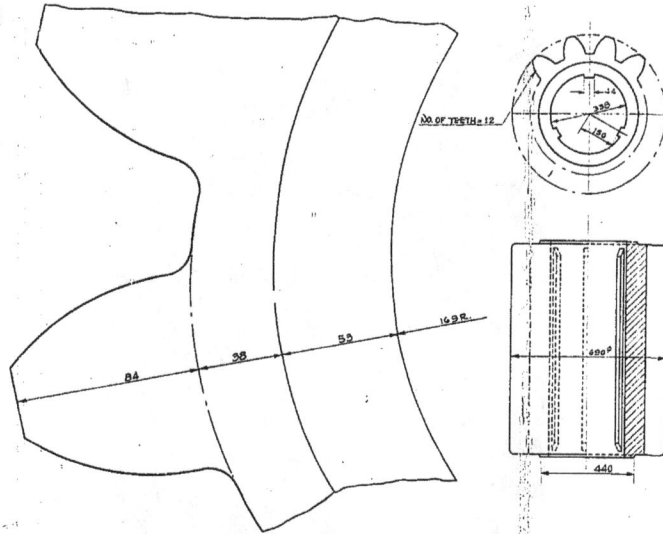

, Figure 42
GUN TURRET TRAINING PINION - STEEL
(Dimension in mm)

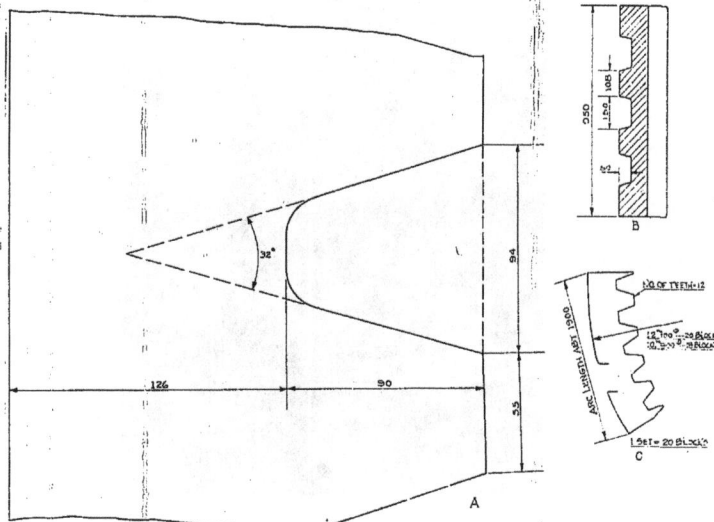

Figure 43
GUN TURRET TRAINING RACK
(Dimension in mm)

44

Figure 44
TRAINING ENGINES

Figure 45
TRAINING ENGINES

45

43 (C) are not reliable, as it could not be verified whether the number of segments of training rack used was 18 or 20.

21. A vertical hydraulic engine of 500 hp (Figures 44 and 45) was built into each side of the turntable. (The entrance to the training engine spaces can be seen at (C) in Figures 55, 57, and 58). Referring to Figure 41 (B), the hydraulic motor drove the inner member (A) of a "coaster" gear through a core clutch at (CC) (clutch is not shown on the sketch). This was named after the bicycle drive which it closely resembled. The inner member (A) drove the outer member (B) frictionally through three (total six - three for each direction) "flattened" rollers (D). (B) was keyed to the first of a train of straight-toothed pinion ending in the training pinion (shown in Figure 42) which drove the turret at two degrees per second. According to the gear ratios shown in Figure 41, this would have required a maximum speed of 53 rpm from the training engine. As this sketch was made from memory, (with the exception of the final driving pinion) these figures could not be checked.

22. It was originally intended to fit the "coaster" gear in conjunction with a hydraulically operated band brake on the driving shaft from the hydraulic motor. The band brake would have acted as a non-reversing mechanism, a function normally performed by the worm, but it was not a success and was not fitted. The training gear, therefore, in its final form was NOT non-reversing, and the "coaster" gear served only as a substitute for the normal friction discs to assist in absorbing the firing thrust when firing an outer gun, or the inertia of the turret on sudden reversal of the direction of training. It is claimed that it absorbed about 50% of the inertia of the target when the direction of training was suddenly reversed at full speed, and the slip under these conditions was 30 minutes of arc. The method of operation of this gear can clearly be seen from Figure 41 without further description.

23. Considerable trouble was experienced in the early stages with severe scoring of the driving face on the outer drum (B). This was overcome by forced lubrication (without unduly increasing the amount of slip). A one hp electric pump was used for this purpose as well as to lubricate all the pinion teeth and bearings. Although the design of the "coaster" gear and band brake was not successful as a non-reversing gear, it was considered a stroke of luck in that the trials had resulted in a neat and compact substitute being found for the normally used bulky friction discs. The overall dimensions of the outer drum are claimed to be about three feet external diameter and eight to nine inches in height. This would indicate a very small area of driving face on the "rollers". Unfortunately no sample of this gear could be found, so these data were not confirmed.

24. <u>Elevating and Slide Locking Gear</u> (Figures 49 and 50.)

The normal cylinder and piston type of elevating gear was used. The main dimensions of the cylinder (taken from a cylinder found at KAMEGAKUBI) are given in Figure 49. Referring to Figure 50 the upper end of the piston rod was connected to a crosshead and slipper (B) running in the slipper guides (C). (These guides can just be seen in Figures 46 and 47, in which they are labeled (D) The crosshead was connected to a piston rod and piston working on the hydraulic shifting cylinder (D) (Figure 50). The position of the piston in the cylinder was controlled by a hand lever on the side of the cradle. It was thus possible to vary the distance between the gun trunnions and the elevating piston rod crosshead (the elevating radius arm). In the "In" position (see Figure 50 (B))the radius arm was a minimum and in this position, the elevating limits of the gun are +45° to -5°. In the "Out" position (see (A) of Figure 50) the radius arm was a maximum, and the elevation of the gun was limited between +41° and +3°. The gun was normally used with the crosshead in this position. The crosshead could be locked at either end of its stroke by

46

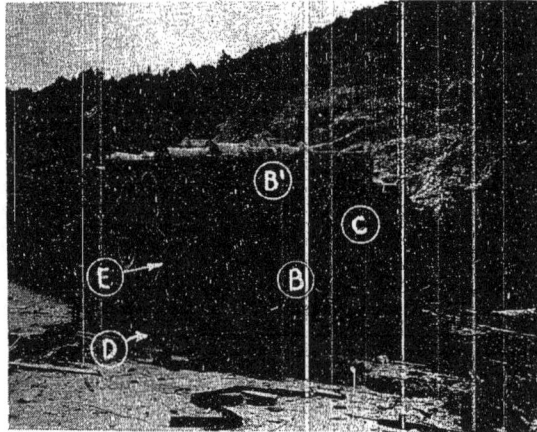

Figure 46
18-INCH SLIDES FOR BB SHINANO

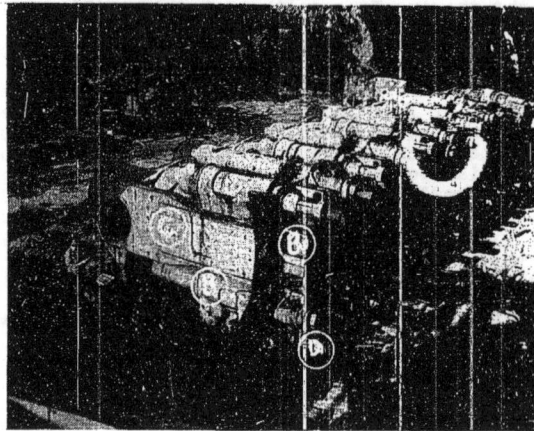

Figure 47
18-INCH SLIDES FOR SHINANO

47

Figure 48
18-INCH SLIDES FOR SHINANO

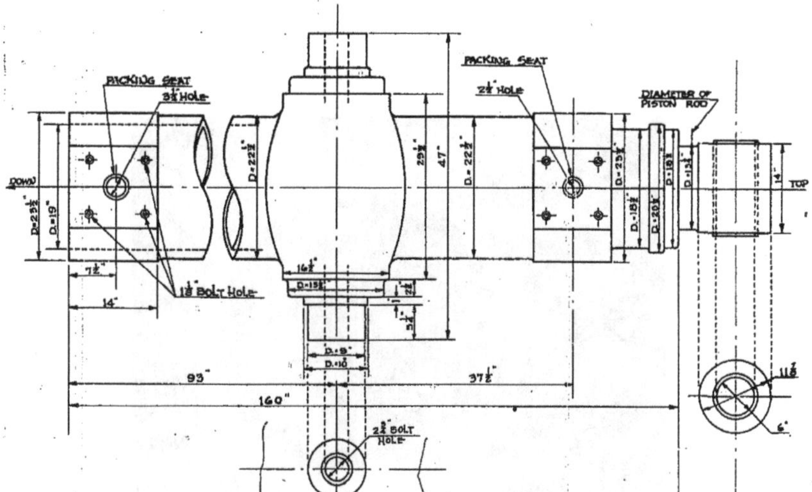

Figure 49
ELEVATING CYLINDER

48

POSITION FOR MAX
DEPRESSION & ELEVATION (-5°~+45°)

POSITION FOR
SLIDE LOCKING

HAND WHEEL FOR LOCKING Ⓔ

STROKE
Ⓐ
-5°
0°
Ⓑ
Ⓐ
B
+3°
Ⓐ
Ⓑ
Ⓒ
Ⓐ
ABT.+4°
A

SHIFTING CYLINDER Ⓓ

GUN. LOCKING PLUG-
BY HAND

MAX. TRAVEL OF ELEVATING PISTON

SECTION "A-A"

ELEVATING CYLINDER

PUSH-IN LIMIT OF ELEVATING PISTON

Ⓕ
C

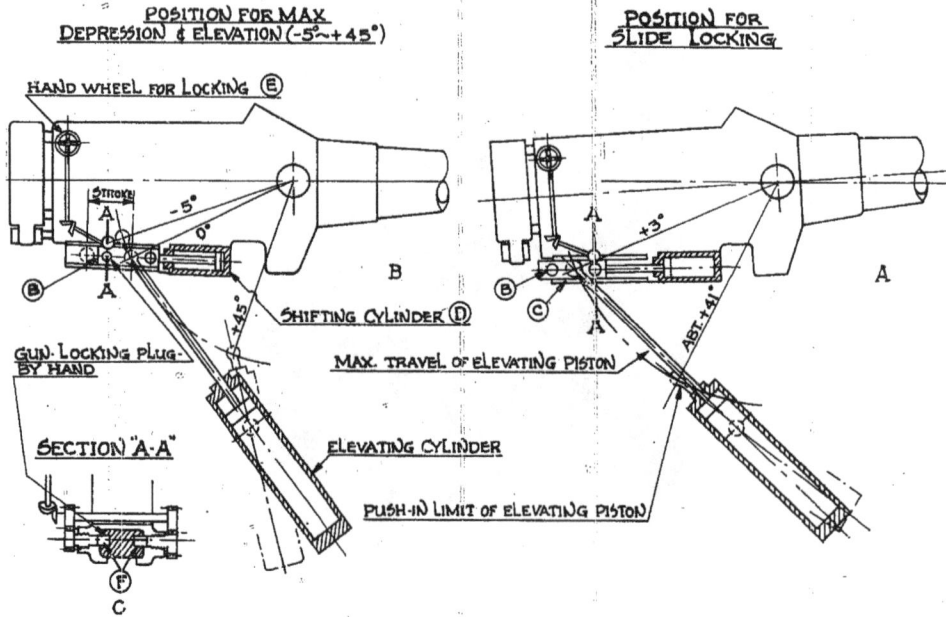

Figure 50
ELEVATING GEAR

Figure 51
18-INCH SLIDE FOR B⁹ SHINANO

49

locking bolts (F) (see (C) of Figure 50) operated by pinion and bevel
gearing from hand wheel (E). (This can also be seen in Figures 47, 48
and 51.)

25. Slide Locking Gear

The object of this movable crosshead, was to act as a form of slide locking gear. With the crosshead locked in the "Out" position, as in Figure
50 (A), it was only necessary, when the gun was at elevation, for the
elevating handwheel to be put hard over to "Depress", and the gun would
automatically be stopped at 3° of elevation by the normal "cut off" gear
coming into operation as the piston arrived at the limit of its travel in
the cylinder. The slide was not locked during the loading operations,
but the elevating handwheel was kept at "Depress", thus keeping the gun
stationary at 3° elevation. A "tell-tale" operated by the rammer operator showed the pointer when the gun was loaded and rammers withdrawn. A
hand operated slide locking bolt was fitted for locking the guns at 1°
elevation when secured for sea.

26. The designed elevating speed was 6°/sec, but it is claimed that in
practice 8°/sec was obtained.

27. Rate of Fire

Estimates given of the loading cycle times of the 18-inch mounts vary
from 40 seconds at full elevation, to 28 seconds at an unknown but pre-
sumably low elevation. When first questioned on this subject, Lt. Comdr.
MATSUMURA stated that a prolonged series of trials had been carried out
to establish the actual rate of fire of these mounts, and the best result
obtained was one round in 28 seconds, but one round in 30 seconds was the
normal rate. The angle of elevation at which these results were obtained
is not known. Mr. SUGIYAMA was then asked the times for the individual
loading operations. He could not remember them exactly, but as far as he
could remember, they were approximately as follows:

```
Depressing gun from 20° elevation to loading
    angle to 8°/sec and allowing for cut-off ... 2.75 - 3 sec
Opening breech .......................................... 2.0 - 2.5 sec
Moving shell bogie forward .................................: 3 sec
Ramming shell ............................................... 3 sec
Withdrawing rammer and returning bogie ..................... 5 sec
Swinging cordite cylinder and rammer to
    loading position ...................................... 3 sec
Ramming charge ............................................. 3 sec
Withdrawing rammer ......................................... 3 sec
Returning cordite cylinder and rammer to
    firing position ...................................... 3 sec
Closing breech ............................................. 2 sec
Elevating gun .......................................... 2.75 - 3 sec
Recoil and run-out ..................................... 2.5 - 3 sec
```

This gave a total time for the loading cycle at 20° elevation, (neglect-
ing time on aim) of 35 seconds to $36\frac{1}{2}$ seconds - say 36 seconds. This
agrees with a loading cycle of 30 seconds at 3° elevation, and approximately 40 seconds at 41° elevation. The rate of fire given by Captain
IWASHIMA and Captain DATE was 1.5 rounds per minute at full elevation,
and it seems reasonable to accept this as the correct rate. The speed of
ammunition hoists, and handling room operations allowed an ample margin
to supply shell and charge in the gunhouse with no delay in firing.

50

28. Turntable (Figures 52 to 59.)

Figure 58 is a diagrammatic representation of the construction of the turntable, which was an entirely riveted, fabricated structure. The weight of the bare turntable as it appears in the plates was approximately 300 tons. In the factory it was lifted and tipped upside down by special lifting gear secured to removable lifting plates ((A) in Figure 54, 55 and 57) for machining the roller path. An idea of its size can better be obtained from the general view of the proving ground at KAMEGAKUBI in Figures 8 and 9. In these photographs it can be compared with other units of the 18-inch mount, and also with a 6-inch mount. The units shown are (A) 18-inch upper and lower cordite handling rooms, (B) 18-inch turntable, (C) 18-inch upper and lower shell handling rooms (D) 6-inch triple gunhouse and working chamber, (E) 16-inch cradle and slide, (F) 18-inch cradles and slides.

Approximate dimensions of the lower roller path are given in Figure 59 (A)

29. Trunnion Brackets (Figures 55 to 57 and 60.)

There were only four trunnion brackets for the three guns. Further description of these trunnion brackets is unnecessary, as full details are given on Figure 60 which is self-explanatory. This sketch and Figure 58 were both made from inspection of the incomplete turntable depicted in Figures 54 to 57.

30. Cradle and Slide (Figures 46 to 48, 51, 59 and 61.)

The main structure of the cradle consisted primarily of two semi-cylindrical steel castings (B) and (B$_1$) joined at (A), Figure 51, the center line of the keyways for the gun keys. The necessary housings for recoil and run-out cylinders and elevating piston rod slipper guides were an integral part of the main castings. (B) and (B$_1$) were joined together by two large side plates (C) in Figure 61. A splinter plate (A) and a light copper chase protection plate (P) were fitted to the front of the cradle.

Figure 52
18-INCH TURNTABLE (TRIAL DESIGN SEMI-COMPLETED)

51

Figure 53
—(SAME AS FIGURE 52)

Figure 54
NEARLY COMPLETED TURNTABLE FOR SHINANO

52

Figure 55
(SAME AS FIGURE 54)

Figure 56
(SAME AS FIGURE 54)
Supports Wasted Away by Typhoon
Weight 300 Tons (Approx.)

53

Figure 57
18-INCH TURNTABLES FOR SHINANO - SEMI-COMPLETE

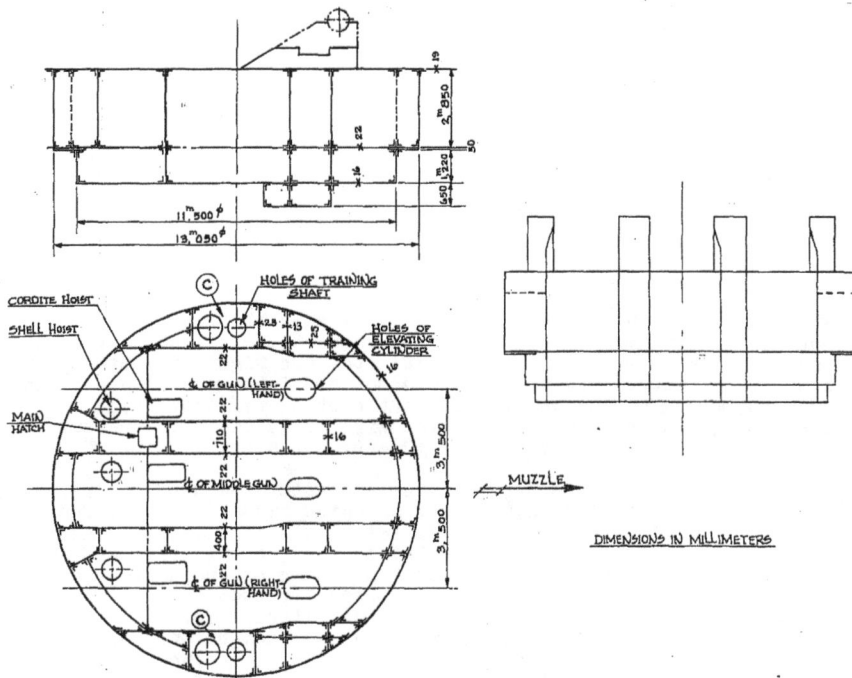

DIMENSIONS IN MILLIMETERS

Figure 58
TURRET TURNTABLE

54

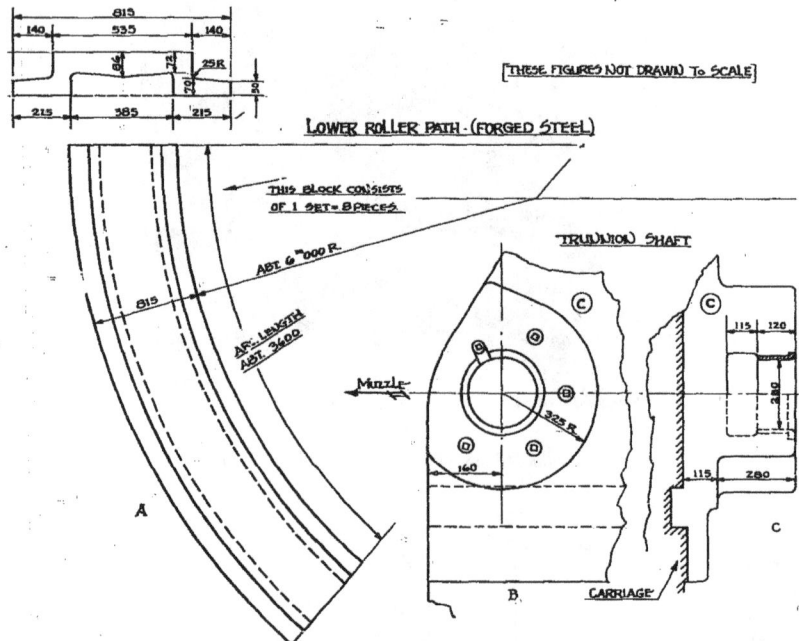

Figure 59
LOWER ROLLER PATH AND TRUNNION SHAFT

31. Recoil and Run-Out

Recoil and run out were controlled by five separate cylinders fitted on top and below the cradle. Figure 51 shows these cylinders from the breech end. They were situated as follows: the top right and bottom left were run-out cylinders, the top left and bottom right were recoil cylinders, and the bottom center was the run-out control cylinder. The run-out cylinders were pneumatically operated with an initial air pressure of approximately 1350 psi and a final pressure of about 2480 psi.

The usual 50-50 mixture of glycerine and water was used on the recoil and run-out control cylinders, neither of which had any unusual feature. The recoil cylinders played only a very small part in the control run-out, and what little effect they had on the run-out speed was produced by a tail on the recoil piston rod working in an extension to the cylinder. It was mainly to avoid weakening the recoil piston and piston rod that an independent run-out control cylinder was fitted. The weight of the cradle and slide was such that only a small balance weight was necessary to balance the gun in elevation. It was to this balance weight (one of which is shown in Figure 10) that the run-out, recoil, and run-out control piston rods were secured. Two cradles and slides, complete with recoil and run-out mechanisms, have been shipped to the United States for use in proving the guns and for detailed examination.

55

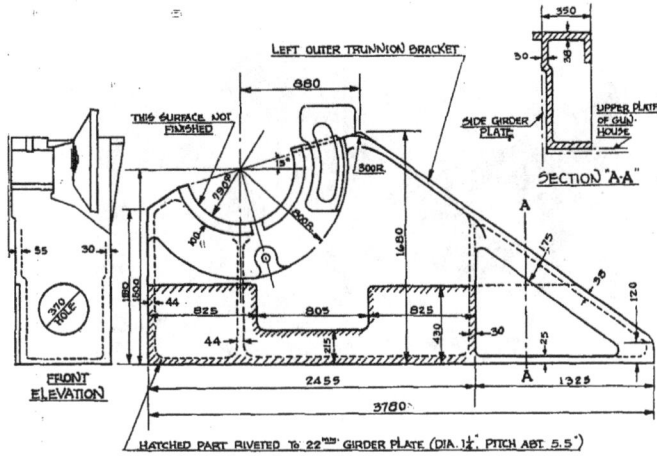

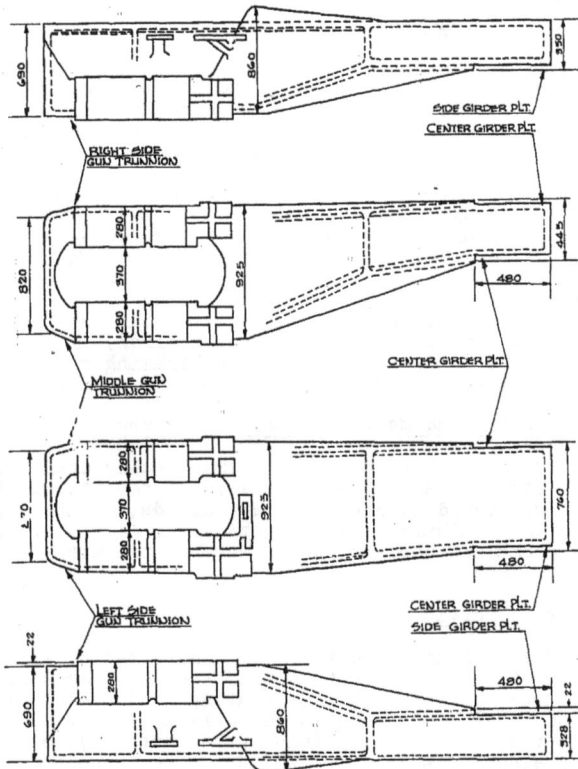

Figure 60
TRUNNION BRACKETS

Figure 61
18-INCH SLIDE FOR SHINANO

Part III

MISCELLANEOUS INFORMATION AND NOTES ON PERFORMANCE OF THE MOUNT IN SERVICE

1.   Secondary Ammunition Supply

Two auxiliary hoists were fitted in each turret. These were of conventional design, using a hydraulic winch, similar to, but smaller than, the
main cordite hoist winch. Both shells and cordite were hoisted vertically in suitable containers. Transfer of shell in the gunhouse was effected
manually by overhead travellers and chain purchase.

2.   With the original design of "Push-Pull" gear secondary supply in the
shell and shell handling rooms was done by overhead traveller and chain
purchase. This was found to be too slow and the "Push-Pull" was modified
to allow the gates to be operated independently of each other, so that
both could be raised at the same time. This provided a clear space between the main girders. Shells were moved in the space by being dragged
and rotated on their bases, by means of a wire fixed at one end, passed
round the shell and then drawn by a hydraulic winch. It was by this means
also that shells were loaded into the upper handling room from the hoist
when embarking or replenishing the supply in this room after an action.

3.   Cordite Drenching and Magazine Flooding Arrangements

The flooding valves for magazines were hydraulically operated by remote
control valves fitted in a cabinet in the lower cordite handling rooms.
The flooding valves could be opened in 20 seconds, and the magazines completely flooded in 15 minutes. The magazine, shell rooms, cordite and
shell handling rooms and cordite hoists were fitted with normal spraying
arrangements.

4.   Performance in Service

The designers of this mount were favorably impressed by its performance
on service. They fully expected that with such a large turret containing
many novel features complaints from operating personnel would be numerous.
This was not so. They admit that it was only used over a period of about
three years and therefore certain inherent defects may not have had time
to develop to a stage at which they would become troublesome. This is

57

correct, but, on the other hand, it is usually found that the first few
years of service of a new type of mount are those when most troubles are
experienced. There was only one serious accident during this time, - and
that was a fatal one. A man was killed in the space between the shell
and shell handling rooms, by being cut in two by a shell bogie. This was
thought to be due to his own carelessness.

5.   The most troublesome feature was the large amount of lubricating oil
used by the cordite hoist racks and winches and by the training gears;
these had begun to show signs of heavy pitting. Complaints were also
made about the noise made by the rollers in the "coaster gear" and diffi-
culty experienced in hoisting shells when the ship was rolling more than
5°, but no further information was available on this. The blast from
these guns was very great and was particularly troublesome in the bridge
area.

6.   Information obtained about the operational use of these guns is as
follows:

    a.   A considerable amount of research has been carried out during
    the past 10 years on the problem of the reduction in size of spreads
    of salvoes in Japanese naval gunnery. In the design and manufacture
    of the 18-inch mounts and, for that matter, all other modern mounts
    great care was taken to reduce to a minimum the backlash in elevat-
    ing and training drives, elevation and training receiver drives, and
    all other sources of large spreads. The normal spreads obtained
    with the 18-inch guns were reported to be about 500 to 600 yards at
    maximum range when firing five and four gun salvoes. When firing
    broadsides the spreads were larger than this. As the solution of
    the problem of reduction in spreads is dealt with fully in NavTechJap
    Report, "Japanese Surface and General Fire Control", Index No. O-31,
    no further details will be given here.

    b.   The 18-inch guns were used against aircraft, mainly against
    torpedo bombers, using Type 3 (incendiary shrapnel) shells. The
    effectiveness of the fire is discussed in some detail in NavTechJap
    Report, "Japanese Projectiles - General Types", Index No. O-19.
    Three widely differing statements were given on the amount of anti-
    aircraft shells carried by YAMATO and MUSASHI and they were (1) 100%
    of full stowage, (2) 40%, (3) 10%. (1) was given by Lt. Comdr.
    MATSUMURA and is probably most unreliable. (2) and (3) were given
    by Captain KURODA and Captain MITSUI, respectively, and it is diffi-
    cult to judge which is the most reliable; probably the 40% figure is
    nearer the correct answer. There was no rapid method of changing
    the type of shell to be fired from these guns other than loading some
    hoists with one type and the remainder with another. Fuzes were set
    in the shell handling rooms, and fuze protectors used to prevent dam-
    age to the fuze before loading. The method of calculating the fuze
    time to be set, is dealt with in NavTechJap Report, "Japanese Surface
    and General Fire Control," Index No. O-31.

ENCLOSURE (A)

LIST OF JAPANESE EQUIPMENT SHIPPED TO THE UNITED STATES

(All for Type 94, 46cm (18") Mount)

NavTechJap Equipment No.	Description	Destination
JE 22-2070 and JE 22-2070 (A)	Left Hand Gun and Breech Mechanism (1)	NPG
JE 22-2071	Right Hand Gun and Breech Mechanism (1)	NPG
JE 50-3180	Cradle and Slide (For left hand gun) (1)	NPG
JE 50-3181	Cradle and Slide (For right hand gun) (1)	NPG
JE 50-3182	Gunhouse Shell Bogie and Rammer (Trial Design) (1)	NRL
JE 50-3183	Gunhouse Shell Bogie and Rammer (Final Design) (1)	NRL
JE 50-3184	Cordite Gunloading Cage (1)	NRL
JE 50-3185	Gunhouse Shell Tilting Bucket and Waiting Tray (1)	NRL

NPG - Naval Proving Ground, Dahlgren, Va.
NRL - Naval Research Laboratory, Anacostia, D.C.

59

www.ingramcontent.com/pod-product-compliance
Lightning Source LLC
Chambersburg PA
CBHW050642150426
42813CB00054B/1155